CAMBRIDGE EARTH SCIENCE SERIES

Editors:
A. H. Cook, W. B. Harland, N. F. Hughes,
A. Putnis, J. G. Sclater and M. R. A. Thomson

Creep of crystals

Creep of crystals

High-temperature deformation processes in metals, ceramics and minerals

JEAN-PAUL POIRIER

Institut de Physique du Globe de Paris
Université Pierre et Marie Curie

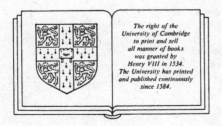

The right of the
University of Cambridge
to print and sell
all manner of books
was granted by
Henry VIII in 1534.
The University has printed
and published continuously
since 1584.

CAMBRIDGE UNIVERSITY PRESS

CAMBRIDGE

LONDON   NEW YORK   NEW ROCHELLE

MELBOURNE   SYDNEY

CAMBRIDGE UNIVERSITY PRESS
Cambridge, New York, Melbourne, Madrid, Cape Town, Singapore, São Paulo

Cambridge University Press
The Edinburgh Building, Cambridge CB2 2RU, UK

Published in the United States of America by Cambridge University Press, New York

www.cambridge.org
Information on this title: www.cambridge.org/9780521261777

First published 1985
Re-issued in this digitally printed version 2005

A catalogue record for this publication is available from the British Library

Library of Congress catalogue card number: 84-11348

ISBN-13 978-0-521-26177-7 hardback
ISBN-10 0-521-26177-5 hardback

ISBN-13 978-0-521-27851-5 paperback
ISBN-10 0-521-27851-1 paperback

To Carol

Creep phenomena have so far been investigated only in terms of old-style technology. For a number of decades, the usual way of tackling similar problems was to apply simple and exact tests (e.g. tensile tests) to the very complex and impure materials used in industry and then to subject the results of these tests to a subtle mathematical analysis. As to the prospects of this way of proceeding, we need only to realize that a piece of iron is far more complicated a structure than, for example, a watch. Now imagine subjecting a watch, without opening it, to a compression test; further trying to draw mathematical conclusion from its undoubtedly very interesting stress–strain curve, and finally dissolving the watch in acids to determine its chemical composition. Although the most accurate experimental tools may be used, and the highest degree of mathematical skill displayed, I doubt whether in this way much valuable information could be obtained about how the watch is working and how it could be improved. A much more promising way is to take the watch to pieces to observe its design and then to study the technological properties of its parts. Translated into terms of our present problem: we must first learn the properties of single crystals, in particular the laws of their plasticity; then we may proceed to a study of polycrystalline metals with more chances of succeeding than hitherto.

E. Orowan, *Trans. AIME*, **131**, 412 (1938).

Contents

Contents

Preface

The physics of high-temperature plastic deformation of solids has recently become an object of interest for earth scientists (structural geologists, tectonicians, physicists of the Earth and planetary interiors) as well as for materials scientists. The reasons for this interest are, however, somewhat different. On the one hand, the materials scientist wants to understand the mechanical behaviour of metals and ceramics to design new materials able to withstand more severe conditions or to process them at a lesser cost of energy and matter. The earth scientist, on the other hand, faced with rocks naturally deformed at large strains or with planetary mantles flowing viscously with characteristic times of millions of years, would like to have a physical basis for the extrapolation of laboratory constitutive equations to inaccessible conditions of strain rate and time, as well as for the diagnosis of past conditions from the present microstructure of deformed minerals. In both cases, the materials (alloys, ceramics or rocks) are often complex, polyphase aggregates whose deformation cannot in general be reduced to that of their simpler constituents. It is, however, impossible to dispense with the important step of understanding the physical processes at play in the deformation of single crystals and single-phase polycrystals.

The present text-book is intended for geologists or geophysicists interested in high-temperature deformation of Earth materials and wishing to get acquainted with the methods of materials science as well as with the most important models and recent experimental results, without getting lost in the jungle of the materials science literature. By the same token, the book is obviously also intended for materials scientists starting in the field of high-temperature deformation. Examples are drawn from the field of metals, ceramics or minerals, as the case may be, in the hope of demonstrating the largely non-specific character of high-temperature deformation.

The reader (senior undergraduate or graduate student) is not required to have any special background other than elementary thermodynamics and knowing what a crystal is; the necessary concepts are introduced at the start and developed as needed (up-to-date, key references are provided) until the reader possesses the tools to understand the physics that lies behind the most elaborate models or subtle controversies (if these are really important!). This book does not claim to give a neutral account of the state of the art (although it is hoped to be dispassionate) and it obviously reflects my own views when universal consensus has not yet been reached.

Chapter 1 gives the indispensable background on mechanical tests. The physical agents of high-temperature plastic deformation are lattice defects: vacancies, dislocations, grain boundaries; these will be introduced in chapter 2. Chapter 3 is devoted to a general exposition of the dependence of steady-state creep rate on temperature and applied stress with the relevant thermodynamic formulation. Chapter 4 deals with recovery-controlled and glide-controlled creep models. The effect of hydrostatic pressure, especially relevant to Earth materials, is treated in chapter 5. Structural modifications (polygonization and dynamic recrystallization) usually accompany high-temperature deformation, and they are often used as a means to determine the palaeo-stresses in naturally deformed rocks; chapter 6 is devoted to these phenomena and to a discussion of their use in geology. Chapter 7 deals with deformation by transport of matter (diffusion creep) and superplastic deformation (involving grain-boundary sliding) and chapter 8 with deformation enhanced by phase transformations (transformation plasticity). Finally, chapter 9 briefly introduces deformation mechanism maps and isomechanical classes.

All chapters start with an abstract and some end with a list of recommended readings. The book ends with a bibliographical list of all papers referred to and with an index.

I warmly thank Andrew Putnis, who persuaded me to write this book, made arrangements for its publication and commented on the manuscript. Mervyn Paterson started everything and I also gratefully remember many delightful talks with him on the subject of creep of crystals. Philippe Gillet read the manuscript and gave the viewpoint of the user; I thank him as I do colleagues and friends who contributed illustration material and preprints.

Paris, October 1983

1
Mechanical background

To understand the physics of high-temperature deformation of
crystals, we first need to describe the rheological behaviour of the
solid in terms of mechanical and physical variables (stress, strain,
temperature, pressure . . .). The description is embodied in constitutive
equations, obtained by means of mechanical tests. In the present
chapter, we summarily introduce the fundamental notions needed:
stress, strain, and the various rheological constitutive equations. At
high temperatures many materials flow viscously and viscous
behaviour is therefore especially important. The principal methods of
mechanical testing – creep at constant stress, deformation at constant
strain-rate and stress relaxation – are presented and compared. The
role of the variables in the constitutive equation is discussed: time, a
special kinematic variable, explicitly appearing in transient creep only;
strain, usually not a good variable, except when it coincides with the
structural variables; strain-rate and stress. Minimum creep-rate,
steady-state creep-rate and constant-structure creep-rate generally
correspond to different conditions and must not be confused. We are
concerned here with uniform deformation, but it may be useful to
consider briefly the criteria for non-uniformity (i.e. localization) of
deformation. Shear localization is a plastic instability manifesting itself
as a stress drop on stress–strain curves.

1.1   Definitions

1.1.1   Stress and strain

To investigate the physics of the deformation processes in crystals
it is first necessary to obtain a description of the phenomenon in terms of the
relevant variables. We are dealing with deformation, i.e. a change of shape
of a crystal or an aggregate of crystals, in response to external forces; this is
obviously a problem in continuum mechanics. The solid must first be
considered as a continuous body, endowed with material properties
represented by constants, dependent on temperature and pressure. The

change of shape and the external forces must be given precise expressions, susceptible of being measured in the course of mechanical tests.

   (i) Let us first consider the external forces. Forces applied to the external surface of a body at rest must clearly be felt inside the body; in other words, a small volume element inside the body is subjected, on its surface, to forces transmitted through the bulk of the body.

   Let us consider (fig. 1.1) a point P inside the body and the three mutually orthogonal planes going through P, normal to the reference coordinate axes Ox_1, Ox_2, Ox_3; these three planes together with a fourth one normal to an arbitrary direction define an infinitesimal tetrahedron. Let $\mathbf{F}_1$ be the force per unit area applied to the face of the tetrahedron normal to Ox_1 at any point P, and let $\sigma_{11}, \sigma_{12}, \sigma_{13}$ be the components of the vector $\mathbf{F}_1$ along the axes Ox_1, Ox_2, Ox_3, respectively. The quantities $\sigma_{21}, \sigma_{22}, \sigma_{23}$ and σ_{31}, σ_{32}, σ_{33} may be defined, in the same way, as the components of $\mathbf{F}_2$ and $\mathbf{F}_3$ applied to the faces of the tetrahedron normal to Ox_2 and Ox_3.

   The vector $\mathbf{F}_1$ is called the *stress vector* on plane Px_2x_3, normal to Ox_1, at point P. If we see plane Px_2x_3 as dividing the body into a positive part (on the side of Px_1) and a negative part, we can say that the stress vector $\mathbf{F}_1$ represents the action of the negative part on the positive part of the solid at

Fig. 1.1. Stresses: The faces of an elementary tetrahedron isolated inside a solid are subjected to resultant forces (per unit area) from the rest of the solid (stress vectors). The stresses are the nine components parallel to the coordinate axes of the stress vectors acting on the faces parallel to the coordinate planes.

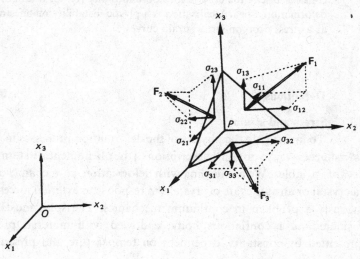

P. The components σ_{ij} along the j axis ($j=1, 2, 3$) of the stress vector relative to the plane normal to the i axis ($i=1, 2, 3$) are called the stress components or the *stresses* for short. The nine σ_{ij} components define the *stress tensor* at point P:

$$\begin{pmatrix} \sigma_{11} & \sigma_{12} & \sigma_{13} \\ \sigma_{21} & \sigma_{22} & \sigma_{23} \\ \sigma_{31} & \sigma_{32} & \sigma_{33} \end{pmatrix} \tag{1.1}$$

At the outer surface of the body the state of stress is imposed by the applied external forces.

It can be shown that in most cases (when there is no body torque) the stress tensor is symmetrical, i.e. $\sigma_{ij} = \sigma_{ji}$, the number of independent stress components being reduced to six.

The stress tensor defines entirely the state of stress at point P in the body, i.e. the stress vector **T** on any plane going through P and normal to an arbitrary direction **n** (n_1, n_2, n_3) can be determined knowing all the σ_{ij}. It follows from the equilibrium condition of an infinitesimal tetrahedron whose fourth face is normal to **n**, that:

▶ $$T_i = \sum_j \sigma_{ij} n_j \quad (i, j = 1, 2, 3) \tag{1.2}$$

Coming back to the table of the stress components, we can distinguish the diagonal components σ_{ii} called *normal stresses* and the off-diagonal ones σ_{ij} ($i \neq j$) called the tangential or *shear stresses*. By an appropriate change of coordinates a given stress tensor may always be put under a diagonal form, where the only non-zero terms are the diagonal ones $\sigma_1, \sigma_2, \sigma_3$ called the *principal stresses*.

The sum of all diagonal terms is independent of the coordinate system; it is called the *trace* of the tensor. The *hydrostatic pressure* is equal, by definition, to one third of the trace:

▶ $$P = \tfrac{1}{3}(\sigma_{11} + \sigma_{22} + \sigma_{33}) = \tfrac{1}{3}(\sigma_1 + \sigma_2 + \sigma_3) \tag{1.3}$$

It is always possible to decompose a given stress tensor into a hydrostatic part and a *deviatoric* part with a trace equal to zero:

$$\begin{pmatrix} \sigma_{11} & \sigma_{12} & \sigma_{13} \\ \sigma_{21} & \sigma_{22} & \sigma_{23} \\ \sigma_{31} & \sigma_{32} & \sigma_{33} \end{pmatrix} = \begin{pmatrix} P & 0 & 0 \\ 0 & P & 0 \\ 0 & 0 & P \end{pmatrix} + \begin{pmatrix} \sigma_{11}-P & 0 & 0 \\ 0 & \sigma_{22}-P & 0 \\ 0 & 0 & \sigma_{33}-P \end{pmatrix} \tag{1.4}$$

(ii) Let us now move to a definition of the deformation. Under the action of the applied forces, a point of coordinates x_i in the body moves by a *displacement* vector $\mathbf{u}(x_i)$. If the displacement vector is the same for all the points of the body we have a rigid-body translation. If the displacement inside the body is not uniform, i.e. if we have a *gradient* of *displacement* between two neighbouring points, the distance between them is altered, with the consequence that a small volume element, say a cube, will suffer a change in dimensions and shape (and become a parallelepiped). A measure of these changes is embodied in the gradient-of-displacement tensor, whose nine components are the values of $\partial u_i/\partial x_j$ ($i, j = 1, 2, 3$). Better still, the symmetrical part of the gradient-of-displacement tensor can be directly related to the change in the element of length, if the displacement gradients are infinitesimal. We will define the infinitesimal *strain tensor* ε_{ij} as the symmetrical part of the displacement-gradient tensor and use it as a measure of the deformation:

$$\blacktriangleright \qquad \varepsilon_{ij} = \frac{1}{2}\left(\frac{\partial u_i}{\partial x_j} + \frac{\partial u_j}{\partial x_i}\right) \tag{1.5}$$

By definition, the strain tensor is symmetrical. The diagonal terms ε_{ii} are called the *stretches* (positive or negative) and represent the relative length changes of segments parallel to the coordinate axes; the off-diagonal terms are called the *shear strains* and are related to the shape change. The physical meaning of the ε_{ij} is visualized in fig. 1.2 in the case of a two-dimensional deformation: a surface element initially square changes into a parallelo-

Fig. 1.2. Strains (two-dimensional case): Square $OACB$ is deformed into a lozenge $OA'C'B'$. Stretches (ε_{ii}) represent the relative length change of segments parallel to the axis ($\varepsilon_{11} = (OA' - OA)/OA \approx (OA'' - OA)/OA$); shear strains represent the change of shape ($\varepsilon_{12} = \varepsilon_{21} = (\alpha + \beta)/2 = \gamma/2$).

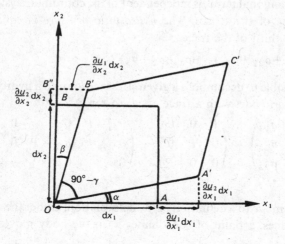

gram and the angle between the Ox_i and Ox_j axes takes the values $90° - \gamma_{ij}$ where γ_{ij} the shear angle is equal to:

$$\gamma_{ij} = 2\varepsilon_{ij} \tag{1.6}$$

As for the stress tensor, we can define the trace, equal here (to first order) to the dilatation or relative volume change:

$$\frac{\Delta V}{V} = \varepsilon_{11} + \varepsilon_{22} + \varepsilon_{33} = \frac{\partial u_1}{\partial x_1} + \frac{\partial u_2}{\partial x_2} + \frac{\partial u_3}{\partial x_3} \tag{1.7}$$

The strain tensor can be decomposed into a diagonal part corresponding to a dilatation without shape change and a part with zero trace corresponding to shape change at constant volume.

1.1.2 Rheology

Rheology is the branch of physics closest to mechanics. It gives a phenomenological account of the mechanical behaviour of matter which involves its material properties: the free fall of a ball of putty, a ball of steel or the contents of a glass of water is described in an identical manner by classical mechanics, but the behaviour of these materials is entirely different when they reach the ground; it can be described by *constitutive equations* which, in addition to continuum mechanics parameters, like stress or strain, involve '*material' parameters*, characteristic of the materials. The material parameters depend on temperature, pressure and the microstructure at all scales of the materials, but rheology considers them only as phenomenological constants and does not concern itself with the microscopic physics that determines them. It is the purpose of this book to examine the physical processes underlying the high-temperature rheological behaviour of materials and it obviously must be described first. It is convenient to define types of rheological behaviour – elasticity, plasticity, viscosity – and we will briefly review their characteristics and constitutive equations. Although many complex materials (polymers, foodstuffs, paints, etc.) exhibit properties which participate of several types simultaneously, crystals are relatively simple and their behaviour can usually be described in the framework of one type.

(i) Elasticity

The elastic behaviour is particular in that it is the only one that is thermodynamically reversible. An elastic body subjected to a state of stress σ_{ij} instantaneously acquires strains ε_{ij}, and elastic strain energy is stored in the body; upon release of the stresses, the strains instantaneously disappear (fig. 1.3(*a*)) and the stored energy is integrally restored without any dissipation into heat (for a perfectly elastic body).

The constitutive relation (Hooke's law) is then a true mechanical equation of state relative to well-defined thermodynamic states, independent of the path followed to reach them. It linearly relates the stress tensor to the resulting strain tensor:

Fig. 1.3. Rheological behaviours: Strain ε is represented as a surface as a function of stress and time $\varepsilon = f(\sigma, t)$. The paths of stress $\sigma(t)$ and strain $\varepsilon(t)$ are represented by simple and double arrows respectively. (a) Elasticity: $\varepsilon = \sigma/\mu$, ε is time independent. (b) Newtonian viscosity: $\varepsilon = (\sigma/\eta)t$ or $\dot{\varepsilon} = \sigma/\eta$, ε varies linearly with time at a rate proportional to stress. (c) Bingham solid: for stresses higher than a threshold stress σ_T, the behaviour is viscous, not necessarily linear. (d) Rigid, perfectly plastic solid: for stresses lower than σ_P, $\varepsilon = 0$, the stress cannot be higher than σ_P and $\sigma = \sigma_P$ is strain-independent. ε_P is the permanent plastic strain at time t. (e) Rigid plastic solid with hardening: $\sigma = \sigma_P$ increases with ε.

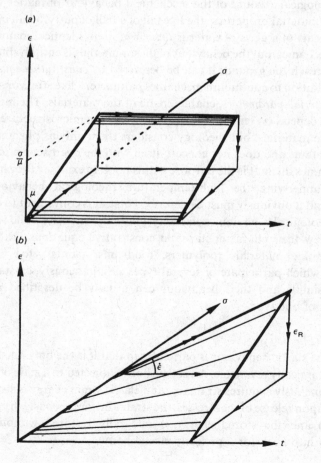

(c)

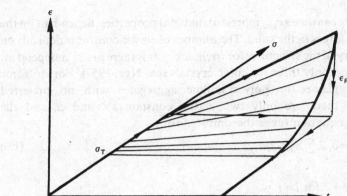

(d)

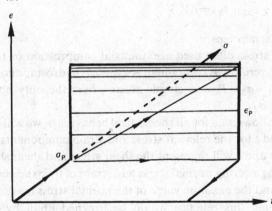

(e)

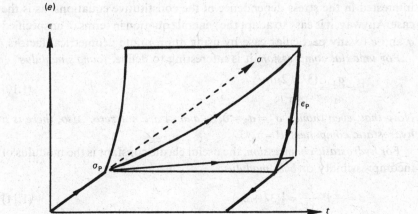

▶ $$\sigma_{ij} = \sum_{ij} c_{ijkl} \varepsilon_{kl} \tag{1.8}$$

The elastic constants c_{ijkl} represent material properties, dependent on the interatomic forces in the solid. The number of elastic constants depends on the symmetry of the structure: for triclinic crystals there are 21 independent constants and only three for cubic crystals (see Nye, 1957). For isotropic solids, like glasses or polycrystalline aggregates with no preferred orientation, there are only two elastic constants, λ and μ, and the constitutive equation takes the simpler form:

▶ $$\sigma_{ij} = \delta_{ij} \lambda \sum_{i} \varepsilon_{ii} + 2\mu \varepsilon_{ij} \tag{1.9}$$

where

$$\delta_{ij} = 1 \quad \text{for } i = j$$

$$\delta_{ij} = 0 \quad \text{for } i \neq j$$

$$\sum_{i} \varepsilon_{ii} = \varepsilon_{11} + \varepsilon_{22} + \varepsilon_{33} = \Delta V / V$$

is the trace of the strain tensor.

Simple states of stress often used are: uniaxial compression or tension, where the only non-zero stress component is $\sigma_{11} = \sigma_1$, hydrostatic compression where $\sigma_1 = \sigma_2 = \sigma_3 = P$, and simple shear where the only non-zero component is, say, σ_{23}.

N.B. In all that follows, and for all rheological behaviours, we will use the plain symbols σ and ε for the relevant stress and strain components in any state of stress, i.e. σ and ε will represent the shear stress and shear strain of simple shear, the engineering normal stress and strain of uniaxial compression or extension and the absolute value of the normal stress and strain of pure shear. Although this practice would be frowned upon by people trained in mechanics, it has no harmful consequences if we are mostly interested in the stress dependence of the constitutive equations, as is the case. Anyway, it is easy to adapt the general equation in terms of unspecified σ and ε to any particular case by using appropriate geometrical factors.

For uniaxial compression, it is interesting to define *Young's modulus*

$$E = \frac{\sigma_1}{\varepsilon_1} = \frac{(3\lambda + 2\mu)\mu}{\lambda + \mu} \tag{1.10}$$

Note that, even though $\sigma_2 = \sigma_3 = 0$, ε_2 and ε_3 are not zero. Also, there is a hydrostatic component $P = \sigma_1/3$.

For hydrostatic compression, the useful elastic constant is the modulus of incompressibility or *bulk modulus*:

$$B = -P \frac{V}{\Delta V} = \tfrac{1}{3}(3\lambda + 2\mu) \tag{1.11}$$

For simple shear, the relevant modulus is the shear modulus μ. Hooke's law takes then the simple form:

$$\blacktriangleright \qquad \sigma = 2\mu\varepsilon \qquad\qquad (1.12)$$

where σ and ε are the only non-zero components of stress and strain.

As the elastic constants depend on the interatomic forces in the crystal, they mildly depend on temperature and pressure. This anharmonic effect causes a departure from linearity in Hooke's law, negligible in most experimental conditions, but which must be taken into account for the high pressures existing deep inside the Earth.

The elastic constants have the dimension of stresses and are measured in pascals (newton/m^2) or bars (1 MPa = 10 bars). Typical values for μ range between 100 kbars and 1 Mbar, for most materials.

Table 1.1 gives values of μ, B and their pressure and temperature derivatives, when known, for selected materials. Note that all single crystals are elastically anisotropic and that the shear modulus corresponds to different elastic constants (or linear combinations of them) for various orientations of the shear plane and direction with respect to the crystal

Table 1.1

Crystal		Density	B	$\left(\dfrac{dB}{dP}\right)_T$	$\left(\dfrac{dB}{dT}\right)_P$	μ	$\left(\dfrac{d\mu}{dP}\right)_T$	$\left(\dfrac{d\mu}{dT}\right)_P$
Aluminium	Al	2.70	0.82	4.4	−0.16	0.26	1.8	−0.1
Copper	Cu	8.92	1.33	5.4		0.5	1.7	−0.2
Gold	Au	19.30	1.6	6.1		0.3	1.3	−0.1
Iron	Fe	7.86	1.7	5.9		0.8	2	−0.2
Lead	Pb	11.34	0.42	5.5		0.1	1.3	−0.1
Magnesium	Mg	1.74	0.32	3.9		0.17	1.3	−0.1
ε-Cobalt	Co	8.90	1.83	4.3		0.58	1.8	−0.27
Silicon	Si	2.34	0.98	4.2		0.7	0.5	−0.05
Halite	NaCl	2.49	0.14	5.4	0.14	0.26	2.1	0.05
Periclase	MgO	3.60	1.62	4.2	−0.15	1.30	2.4	−0.2
Rutile	TiO$_2$	4.26	2.1	6.6	−0.5	1.1	1.8	−0.3
Calcite	CaCO$_3$	2.71	0.69			0.37		
Quartz	SiO$_2$	2.65	0.3	6.5		0.47	0.3	
Corundum	Al$_2$O$_3$	3.97	2.4	4.3		1.99	1.6	
Forsterite	Mg$_2$SiO$_4$	3.22	1.29	5.3	−0.15	0.81	1.8	−0.13
Garnet	(Mg, Fe)$_3$Al$_2$Si$_3$O$_{12}$	4.1	1.5	5.4	−0.2	0.9	1.4	−0.1
Ice	H$_2$O	0.90	0.073			0.025		

The bulk modulus B and the shear modulus μ are given in Mbar, their pressure and temperature derivatives are respectively dimensionless and given in kbar/°C. All values are reduced to atmospheric pressure (1 bar) and room temperature.

lattice: for instance in cubic crystals the modulus relevant to shear on (001) along [100] is c_{44},[†] whereas it is $\frac{1}{2}(c_{11} - c_{12})$ for shear on (110) along [1$\bar{1}$0]. However, the elastic anisotropy is usually rather small and taking it into account will seldom drastically change things.

(ii) *Viscosity*

Viscous behaviour is essentially dissipative, hence irreversible. In the simplest case of the viscous fluid, an instantaneously applied shear stress, even extremely small, causes the fluid to flow, i.e. to deform with time at a rate measured by the shear strain-rate $\dot{\varepsilon} \equiv d\varepsilon/dt$, dissipating power at the rate $\sigma\dot{\varepsilon}$; in the case of the linear viscous fluid the resulting strain-rate is proportional to the applied stress (fig. 1.3(b)):

▶ $$\sigma = \eta\dot{\varepsilon} \tag{1.13}$$

The viscosity η is a material constant, and in the case of the linear viscous fluid it is independent of the stress (*Newtonian viscosity*). In many instances, the viscosity is non-Newtonian and depends on the stress (usually decreasing as the stress increases), and the strain-rate–stress relationship is no longer linear. If the stress is removed after a given time there remains a permanent strain $\varepsilon = (\sigma/\eta)t$; all of the mechanical energy has been dissipated into heat.

If the flow starts only if the stress is greater than a threshold value, the behaviour is that of a *Bingham solid*.

$$\left.\begin{array}{l} \sigma < \sigma_{\mathrm{T}}: \ \dot{\varepsilon} = 0 \\ \sigma \geqslant \sigma_{\mathrm{T}}: \ \dot{\varepsilon} = (\sigma - \sigma_{\mathrm{T}})/\eta \end{array}\right\} \tag{1.14}$$

The viscosity can be Newtonian or not (fig. 1.3(c)); for stresses much larger than the threshold stress (which may be small) the distinction between a fluid and a solid is somewhat academic: whether or not there was a finite threshold stress, the main point is that the flow rate $\dot{\varepsilon}$ depends on the stress. The flow can be characterized by an *effective viscosity* $\eta = \sigma/\dot{\varepsilon}$ or by the *stress sensitivity of the strain-rate* d log $\dot{\varepsilon}$/d log σ.

Viscosity is expressed in poises (cgs unit or 1 barye s) or in Pa s (1 Pa s = 10 poise): its values for different materials are spread over a considerable range of magnitudes: $\eta = 10^{-2}$ poise for water at room temperature and $\eta \approx 10^{22}$ poise for the rocks of the Earth's mantle. It is usually strongly temperature dependent and decreases as temperature increases. As we will

---

† The stress and strain tensors being symmetrical, the indices i and j and k and l can be interchanged in c_{ijkl}. The elastic constants are normally writen c_{mn} where m and n stand for the pairs ij and kl with the following convention: Indices m or n: 1, 2, 3, 4, 5, 6 respectively correspond to pairs ij or kl: 11, 22, 33, 23 or 32, 13 or 31, 12 or 21. Thus, for example, $c_{12} = c_{1122}$ and $c_{44} = c_{2323}$.

see later, high-temperature creep of solids under constant stress can be considered as a viscous flow process and it is one of the purposes of the present book to investigate the elementary physical mechanisms responsible for the viscosity and its dependence on various parameters. The microscopic flow processes have a spectrum of relaxation times usually dominated by one characteristic time τ. *A given material may appear to flow as a viscous fluid on a time scale longer than τ, whereas it will behave as a rigid (or elastic) solid on shorter time scales.* To express this fact, Reiner (1969) introduced the Deborah number,[†] D, ratio of the time of relaxation to the time of observation t_0:

$$D = \frac{\tau}{t_0} \tag{1.15}$$

For a very small D, the material behaves as a fluid and for a very large D it behaves as a solid.

It is easy to calculate a global value of τ in the case of *viscoelastic bodies* which have viscous and elastic properties (most materials are in this category) represented by the viscosity η and the shear modulus μ. Analogue mechanical systems can be built using springs and dashpots.

The Maxwell body is represented by a spring and a dashpot in series (fig. 1.4(a)); the total strain-rate is the sum of the elastic and viscous strain-rates and the constitutive equation is:

▶
$$\frac{d\varepsilon}{dt} = \frac{1}{\mu}\frac{d\sigma}{dt} + \frac{\sigma}{\eta} \tag{1.16}$$

The Kelvin–Voigt body is represented by a spring and a dashpot in parallel (fig. 1.4(b)); here the stresses are additive and the constitutive equation is:

▶
$$\sigma = \mu\varepsilon + \eta\,\frac{d\varepsilon}{dt} \tag{1.17}$$

If a constant strain ε_0 is suddenly imposed on a Maxwell body, the stress relaxes from the elastic values $\mu\varepsilon_0$, to zero for infinite time (fig. 1.4(a)):

$$\sigma = \mu\varepsilon_0 \exp\left(-\frac{t}{\tau}\right) \tag{1.18}$$

If a constant stress σ_0 is suddenly imposed on a Kelvin–Voigt body, the strain increases from zero to the elastic value σ_0/μ for infinite time (fig. 1.4(b)):

$$\varepsilon = \frac{\sigma_0}{\mu}\left[1 - \exp\left(-\frac{t}{\tau}\right)\right] \tag{1.19}$$

† After the prophetess Deborah, who prophesied that the mountains would flow before the Lord (Judges 5.5).

For both cases the *relaxation time* is the same:

$$\tau = \frac{\eta}{\mu} \qquad (1.20)$$

The rocks of the Earth's mantle have a viscosity of the order of 10^{21} Pa s and a shear modulus of the order of 10^{11} Pa; τ is then of the order of 10^{10} s (300 years); it is therefore obvious that the mantle transmits as an elastic body seismic waves with periods of a few seconds and flows as a viscous fluid on a time scale of millions of years. A ball of polydimethyl siloxane (otherwise known as silly putty) bounces elastically better than a steel ball if thrown, but if at rest on a table it flows under its own weight, and turns into a puddle.

(iii) *Plasticity*

A *perfectly plastic* solid suffers no permanent deformation if the applied stress is lower than a threshold stress called limit of plasticity or yield stress, σ_P. When the stress reaches σ_P it remains constant at that value while a permanent strain which can take any value appears (fig. 1.3(d)). In

Fig. 1.4. Viscoelastic bodies can be represented by various arrays of elastic springs (shear modulus μ) and viscous dashpots (viscosity η), with a relaxation time τ. (a) Maxwell body: spring and dashpot in series. For an imposed strain, the stress relaxes exponentially to zero. (b) Kelvin–Voigt body: spring and dashpot in parallel. For an imposed stress, the strain exponentially reaches its elastic value.

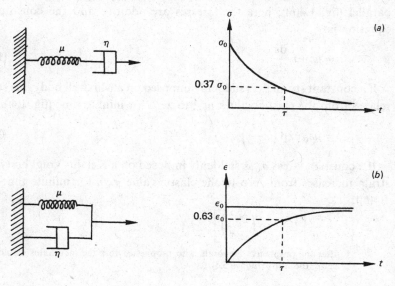

the case of a plastic solid with *work-hardening*, the applied stress must be increased above the yield stress to obtain larger and larger strains (fig. 1.3(*d*)).

Plastic deformation is a shear strain at constant volume that can be caused only by shear stresses; hydrostatic pressure cannot cause permanent strains in non-porous solids.

Plastic deformation is a dissipative, irreversible process, but time does not appear explicitly in the constitutive equations:

$$\left.\begin{array}{l} \sigma < \sigma_P: \ \varepsilon = 0 \\ \sigma \geqslant \sigma_P: \ \varepsilon = f(\sigma) \end{array}\right\} \tag{1.21}$$

However, in practice, plastic deformation is achieved in conditions where plastic strain is imposed at constant strain-rate $\dot{\varepsilon}$ (see § 1.1.3) and the stress necessary to achieve the strain is determined by the properties of the material, so that, in effect, a *perfectly plastic solid, strained at constant $\dot{\varepsilon}$, flows under constant stress exactly as a linear Bingham solid*. A single crystal, subjected to shear stresses, deforms elastically at first; if the applied stress becomes higher than the limit of plasticity (or equivalently, elastic limit) the crystal yields plastically by slip (or glide) over a dense crystallographic plane in a dense direction. A *slip system* is denoted by the Miller indices of the slip plane and the slip direction: $(hkl)[uvw]$. Slip on one given system in crystals of one material is characterized by the limit of plasticity on that slip system, which must be expressed by the shear stress acting on a slip plane along the slip direction. In uniaxial tests (see below) the limit of plasticity is usually measured by the axial yield stress, which varies according to the orientation of the sample with respect to the axial stress. Schmid & Boas (1950) showed that yield always occurred for the same value of the resolved shear stress on the slip plane in the slip direction. The plastic limit for single crystals is therefore called the *critical resolved shear stress* or CRSS. The plastic limit for polycrystals depends on the possible slip systems in the grains and on the preferred orientation or 'texture' of the grains in the sample. According to the active microscopic processes the CRSS may or may not depend on temperature and strain-rate. If it does, and it usually does at high temperatures, it is again possible to describe the behaviour of crystals by using Deborah's number: for a given relaxation time (i.e. for a given T, if $\dot{\varepsilon}$ is high, i.e. if t_0 in (1.15) is small, the stress does not relax and can go to high values before the crystal yields; if $\dot{\varepsilon}$ is small, the stress cannot increase and rapidly relaxes and it is possible to obtain an almost perfect plastic behaviour at low yield stresses, equivalent to a viscous behaviour (fig. 1.5). Similarly, for a given strain-rate it is often possible to obtain an almost constant stress (no work-hardening) at high temperatures (fig. 1.5(*a*) and (*b*)). It is, indeed, the appearance of this behaviour which serves to define operationally the domain of high temperatures. *For laboratory strain-rates, the (fuzzy) limit between low and high temperatures is usually between* $0.3T_m$ *and* $0.8T_m$, where T_m is the absolute melting temperature, depending on the materials; it is generally lower for metals and higher for minerals.

1.1.3   Mechanical tests

Mechanical tests at various temperatures provide the experimental data which must constrain the physical interpretations of the rheological behaviour. Despite their importance, it is out of the scope of this book to give a technical description of even the most usual tests and we will only review them summarily.

Fig. 1.5. Stress–strain curves at imposed strain-rate and constant temperatures for various materials. Note that for large strains and high temperatures strain-hardening vanishes (steady state). (*a*) Silicon steel (after Immarigeon & Jonas, 1974). (*b*) Carrara marble (polycrystalline calcite) (after Schmid *et al.*, 1980) deformed under 3 kbar confining pressure. (*c*) MgO 1.8% Al_2O_3 spinel single crystals (after Duclos, 1981) deformed in compression along $\langle 110 \rangle$ axis ($\dot{\varepsilon} = 6 \times 10^{-6}\,\mathrm{s}^{-1}$). Note yield point at lower temperatures.

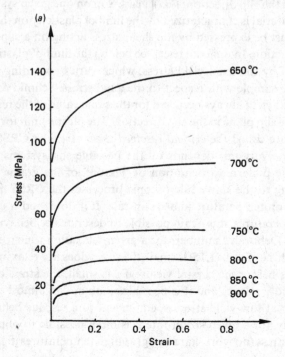

(*a*)

(b)

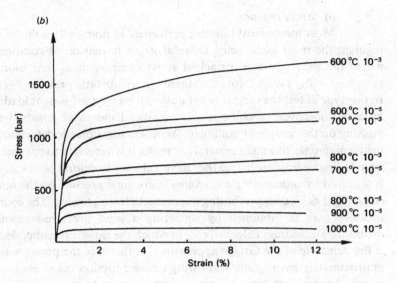

(c)

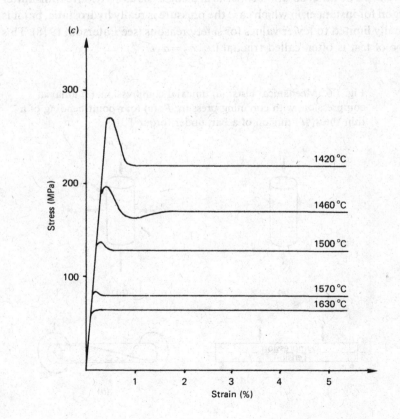

(i) *Stress regimes*

Most mechanical tests are performed in nominally uniform stress regimes, the most usual being *uniaxial stress*, in tension or compression, where the only non-zero principal stress component σ_1 acts along the sample axis (fig. 1.6(*a*)). (Note that there is a hydrostatic pressure $P = \sigma_1/3$.) In this type of test the sample is put under stress by dead-weight loading or by a motor-driven machine whose crosshead moves at constant speed pushing on the sample or pulling it. To obtain some plastic deformation in brittle materials, like most minerals or rocks, it is necessary to prevent them from failing by fracture before they have a chance to deform plastically; this is achieved by superposing a *confining hydrostatic pressure* on the uniaxial stress (fig. 1.6(*b*)), thus preventing microcracks from growing. The confining pressure may be obtained by squeezing a solid pressure-transmitting medium (talc, sodium chloride, etc.) in which the sample is embedded; this is the principle of the Griggs apparatus – in this case the pressure is only approximately hydrostatic and owing to solid friction the stress is rather poorly known, but the confining pressures may be rather high (up to 20 kbar). The other solution consists in using a gas as a pressure transmitter (argon for instance), in which case the pressure is really hydrostatic, but it is usually limited to lower values for safety reasons (see Paterson, 1978). This type of test is often called triaxial ($\sigma_1, \sigma_2 = \sigma_3$).

Fig. 1.6. Mechanical tests: (*a*) uniaxial compression, (*b*) uniaxial compression, with confining pressure P, (*c*) four-point bending of a thin sheet, (*d*) torsion of a bar, under torque Γ.

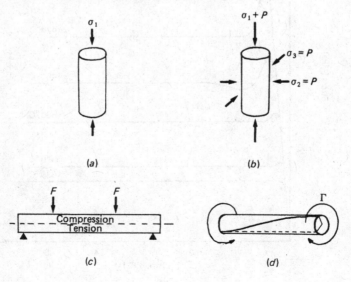

It is common in rock mechanics to use $\sigma_1 - \sigma_3$ as a measure of the applied shear stress and to refer to it by the unfortunate appellation 'differential stress'.

Mechanical tests in conditions other than uniaxial or triaxial are little used and only when no other test is practicable: they involve non-uniform stress states and their interpretation is accordingly more difficult.

Bending tests (fig. 1.6(*c*)) are sometimes used when the material is brittle (i.e. fails at small strains) and available in thin strips only (e.g. carbides). The sample rests on two knife-edges and is loaded in its middle (three-point bending) or in two symmetrically located points (four-point bending). For thin specimens a measurable deflection can be observed even at very small strains, but the stresses are compressive or extensive on opposite sides of the neutral plane, which complicates the analysis.

Impression creep is an indentation creep test in which the indenter is a rod with a flat end; tests on several materials give results comparable with those of conventional uniaxial creep tests (Chu & Li, 1977; Yu & Li, 1977).

Samples in the shape of *helical springs* may be loaded axially in compression or tension. The advantage of this kind of test is that the tensile strains involved are very small for a large displacement of the ends. The shear stresses (maximum at the surface) and shear strains may be calculated for a given compression or extension of the spring (see Reiner, 1969). This test has been used for quartz samples machined from single crystals (Ayensu & Ashbee, 1977) as well as for metal wire springs hanging freely under their own weight (Crossland *et al.*, 1973).

For ductile materials at high temperature, *torsion tests* performed on rodlike samples (fig. 1.6(*d*)) have the interest of allowing stable deformation to large strains (several hundred per cent). The shear stresses and strains are very non-uniform, increasing from zero at the axis to a maximum value at the surface; however, it is often possible to relate the macroscopic torque to the angular velocity, in the same way as the axial stress is related to the strain-rate in uniaxial tests (Rossard & Blain, 1958).

(ii) *Uniaxial tests*

Uniaxial tests differ essentially by the way the sample is stressed: we can distinguish three main varieties, giving complementary information on the rheological behaviour of the materials.

(*a*) *Creep tests.* The sample is dead-weight loaded in tension or in compression and its length is continuously measured with time. The load can be continuously or periodically adjusted to take into account the variation in cross-section of the sample as its length varies at constant volume, and so keep the value of the axial stress constant. The first modern creep experiments under constant stress were done by Trouton & Rankine (1904) on lead. Andrade (1910, 1914), a pioneer in creep research, investigated the behaviour of many metals (Pb, Sn, Fe, Cu, Hg, etc.) at various stresses and temperatures. He found that after a transient, 'even for large extensions, the rate settles down to constant value if the stress is kept constant'. The results of creep experiments are first obtained as a curve giving the length variation with time which must be transformed into the variation of longitudinal strain (or stretch) with time. We have seen (§ 1.1.1) that the infinitesimal stretch of a segment of length l, aligned with Ox_1 axis (the stress axis), is:

$$d\varepsilon = \frac{dl}{l} \tag{1.22}$$

As the length l varies with time from an initial value l_0 to a final value $l(t)$, we obtain the instantaneous value of the total strain at time t by integration of (1.22) from l_0 to l

$$\varepsilon(t) = \int_{l_0}^{l(t)} \frac{dl}{l} = \ln \frac{l(t)}{l_0} = \ln\left[\frac{l_0 \pm \Delta l(t)}{l_0}\right]$$

or

$$\blacktriangleright \qquad \varepsilon(t) = \ln\left[1 \pm \frac{\Delta l(t)}{l_0}\right] \tag{1.23}$$

where the signs $+$ and $-$ respectively correspond to extension and compression, $\Delta l(t)$ is the total length change at time t and $\Delta l/l_0 = \varepsilon_0$ is the length change relative to the initial length. We see that the instantaneous strain, or *natural strain* or *true strain*, refers to the actual length of the sample at time t and that it differs from the strain ε_0 for large strain (table 1.2); for strain up to a few per cent, however, the difference is negligible and ε_0 may be used.

The slope of the creep curve $\varepsilon(t)$ (fig. 1.7) immediately gives the *creep rate* $\dot{\varepsilon}(t)$. When an apparent steady state has been reached, i.e. when the creep curve can be reasonably assimilated to a straight line over a long enough time, the slope is then constant and represents the value of the 'steady-state' creep-rate $\dot{\varepsilon}(\sigma, T)$ corresponding to the constant stress σ and temperature T. The viscosity is immediately obtained: $\eta = \sigma \dot{\varepsilon}^{-1}$. *It is important to notice that in creep experiments the sample flows, so to speak, at its own pace, instead of being driven at an imposed rate. The result is that even a brittle single crystalline specimen can creep in compression under atmospheric pressures without failing, provided that it does not contain microcracks liable to grow and that enough care is taken in aligning the specimen and applying the stress.* In polycrystals, however, incompatibilities at grain boundaries give rise to internal stresses, which may cause cracks to nucleate. Tensile creep tests usually end by necking (a plastic instability) and failure.

(b) *Constant strain-rate tests.* The sample is held between grips or platens in a 'mechanical testing machine' and extended or compressed as the crosshead is driven up or down at constant speed. The machine drive can be mechanical (motor and screws) or hydraulic or a combination of both. The force exerted is measured by a load cell and recorded as a function of time. The length of the sample varies linearly with time, hence the strain-rate $\dot{\varepsilon}_0$ relative to the initial length is constant, but the true strain-rate is not. Although it is commonly used, the name 'constant strain-rate test' is therefore misleading: the true strain-rate can be considered constant only at small strains (see table 1.2). The force–time curve is usually transformed into a true stress-time/strain curve. The true strain is obtained using eq. (1.23) and the stress is obtained by dividing the applied force F by the cross-section area A of the sample; the variation of the cross-section during deformation must be taken into account for large strains; as plastic deformation occurs at constant volume, Al (where l is the length of the sample), it follows that:

$$A = A_0 \frac{l_0}{l_0 + \Delta l} = \frac{A_0}{1 + \varepsilon_0}$$

Fig. 1.7. Creep curves at constant stress and temperature. Single crystals of sodium chloride deformed in compression along $\langle 100 \rangle$ axis (after Poirier, 1972). $1 - T = 780\ °C$, $\sigma = 50\ g/mm^2$; $2 - T = 780\ °C$, $\sigma = 30\ g/mm^2$; $3 - T = 780\ °C$, $\sigma = 25\ g/mm^2$; $4 - T = 765\ °C$, $\sigma = 10\ g/mm^2$.

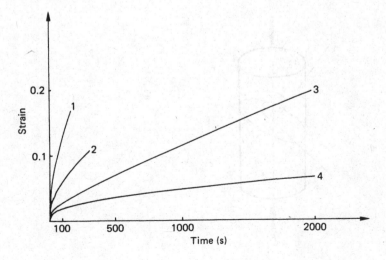

The true normal stress, when strain is uniform, is therefore given by:

$$\sigma = \frac{F}{A_0}(1+\varepsilon_0) \tag{1.24}$$

where A_0 is the initial cross-section area, ε_0 is negative for compression and positive for tension. In the case of single crystals the shear stress σ_R, resolved on the slip plane, in the slip direction, is given by Schmid's formula:

▶ $$\sigma_R = \sigma \cos\psi \cos\lambda \tag{1.25}$$

where ψ is the angle between the load axis and the normal to the slip plane and λ the angle between the load axis and the slip direction (fig. 1.8).

The rheological behaviour of a material can be expressed by a creep curve $\varepsilon(t)$ at $\sigma = $ const or a stress–strain curve $\sigma(\varepsilon)$ at $\dot{\varepsilon} = $ const, and we can now compare these curves (fig. 1.9), keeping in mind that strain-rate appears as the slope of the creep curve and that a variation of σ visible on the stress–

Table 1.2. *Natural strain ε and strain relative to the initial length ε_0* (*compression*)

$\varepsilon_0(\%)$	0.1	1	5	10	15	20	30	40	50
$\varepsilon(\%)$	0.1	1	4.9	9.5	14	18.2	26.2	33.7	40.6

Fig. 1.8. Schmid's formula: In a single crystalline sample, the load axis makes an angle ψ with the normal to the slip plane, and an angle λ with the slip direction. The resolved shear stress is given by $\sigma_R = (F/A_0)\cos\psi \cos\lambda$.

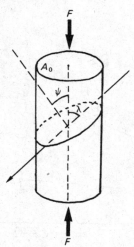

strain curve will correspond to a variation of the derivative of ε on the creep curve: work-hardening is seen on the creep curve as decrease of the slope with time; perfect plastic behaviour corresponds to a linear creep curve (steady state) and work-softening corresponds to an increase of the slope of the creep curve. The phenomenon known as *yield point* sometimes seen on stress-strain curves (fig. 1.9(*d*) and fig. 1.5) is due to a plastic instability: the solid yields for a load which it cannot sustain once the deformation has started, there is a stress drop followed by a region of slow variation of σ and eventually work-hardening appears; the corresponding creep curve is *sigmoidal* and exhibits an inflexion point. Whether a plastic instability in a sample manifests itself as a load drop or not depends on the *hardness* of the machine: when the crosshead moves at constant speed there is a deforma-

Fig. 1.9. Comparison of stress–strain curves ($\dot{\varepsilon} =$ const.) and creep curves ($\sigma =$ const.) for materials with various properties: (*a*) work-hardening, decelerating creep-rate, (*b*) no work-hardening, steady-state creep, (*c*) work-softening, accelerating creep-rate, (*d*) yield point, sigmoidal creep curve.

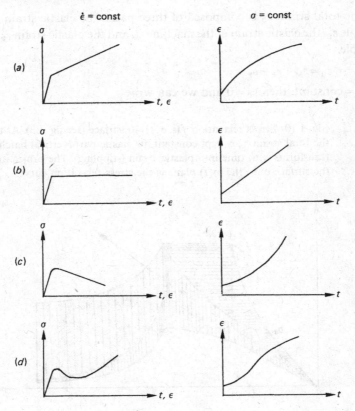

tion not only of the sample but of the whole machine (load cell, screws, etc.) which can be formally described as a spring with a rigidity constant k. If the machine is hard (k large), a rapid variation in length of the sample is immediately felt and the load relaxes visibly; if the machine is soft (k small), the rapid variation in length of the sample is absorbed by the machine and the load drop is ironed out.

(c) *Stress relaxation.* Hard deformation machines may be used for another type of test which can give useful information on the rheology, if properly interpreted. Let us consider the common case of an elasto-plastic material with hardening deformed in compression at constant crosshead speed in a hard machine whose stiffness can be represented by an elastic modulus k (fig. 1.10). Let deformation proceed past the elastic limit into the plastic domain until the flow stress reaches a value σ_0 (for time t_0); at that moment stop the crosshead, i.e. freeze the total strain at its current value ε_0; the stress will decrease with time and would go to zero for infinite time (if the solid is free from internal stresses). The curve $\sigma(t)$ is the stress-relaxation curve.

The total strain ε_0 is composed of three parts: the elastic strain of the sample ε_E, the elastic strain of the machine ε_M and the plastic strain ε_P of the sample.

$$\varepsilon_0 = \varepsilon_M + \varepsilon_E + \varepsilon_P \tag{1.26}$$

If $\varepsilon_0 = $ constant, then $\dot\varepsilon_0 = 0$ and we can write

Fig. 1.10. Stress relaxation $f(\varepsilon, \sigma, t) = 0$ surface (see fig. 1.3). At time t_0, the total strain ε_0 is kept constant, its elastic part (vertical hatching) transforms with time into plastic strain (stippling). The projection of the surface onto the (σ, t) plane is the stress-relaxation curve.

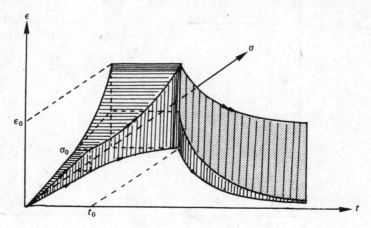

$$\dot{\varepsilon}_M + \dot{\varepsilon}_E + \dot{\varepsilon}_P = 0 \tag{1.27}$$

Now,

$$\varepsilon_M + \varepsilon_E = \left(\frac{1}{k} + \frac{1}{E}\right)\sigma \tag{1.28}$$

where E is Young's modulus of the sample, hence

$$\dot{\varepsilon}_P = -\left(\frac{1}{k} + \frac{1}{E}\right)\frac{d\sigma}{dt} \tag{1.29}$$

This means that, even though the total strain is kept constant, the plastic strain of the sample still increases at a rate proportional to the rate of stress relaxation; what happens is that the elastic strain energy stored in the machine and the sample dissipates irreversibly into plastic deformation of the sample. The total strain is kept constant through an increase in plastic strain and a decrease in elastic strain. With a stiff machine, the relaxation is rapid and in the limit of infinite stiffness only the elasticity of the sample is involved, i.e. *the elastic strain of the sample turns into plastic strain*; stress relaxation is then representative of physical processes taking place in the sample. Conversely, with a soft machine the relaxation is long and is much more representative of the machine than of the sample.

Stress relaxation is used to obtain information on thermally activated processes, as we will see below (chapter 3) (Guiu & Pratt, 1964; Rutter *et al.*, 1978), but we must again emphasize that this method gives physically meaningful results only if the experiment is very well controlled. Essentially:

(i) the machine must be very hard – a hydraulic machine is the best;
(ii) as the relaxation is then very rapid at the beginning, data acquisition must be very fast;
(iii) the test must be rigorously isothermal, which demands excellent temperature control.

(iii) *The samples*
The dimensions and shape are important parameters in mechanical tests and they are not without influence on the creep curve or the stress–strain curve.

(*a*) Polycrystalline specimens obviously must have dimensions at least an order of magnitude greater than the grain size, if the results are to be reproducible. For compression tests, the aspect ratio is also very important: if the sample is too slender it may buckle; if the sample is a flat disk the boundary conditions at the platens become predominant and the strain and the stress become very non-uniform, the hydrostatic pressure increases in the sample as it is squeezed, and the conditions are those obtained in Bridgman's anvils (Jackson & Waxman, 1963).

The best values of the aspect ratio are in the range of 2:1 to 3:1. Absence of lubrication at the platens usually causes barrelling of the sample for large strains hence non-uniformity of strain, but lubrication is not easily achieved at high temperatures as many lubricants do not withstand high temperatures. Besides, the lubricant is squeezed out of the interface, unless it is stored in grooves (Immarigeon & Jonas, 1971).

(*b*) Compression testing of simple crystals is rather delicate, especially when there is one easy slip system. Mechanical tests may give different results for different aspect ratios and orientation of the slip planes with respect to the load axis, according to whether the slip planes meet the free lateral surface or the platens (fig. 1.11(*a*)). In the latter case the strain non-uniformity may be stronger, but in the former case there is also a non-uniformity due to the fact that the slip planes tend to rotate to an orientation normal to the load axis (parallel to it for tensile tests) and that the rotation is impeded at the platens (fig. 1.11(*b*)); this may result in a canting of the specimen and meaningless stress–strain or creep curves. This

Fig. 1.11. Compression tests of single crystals. (*a*) Left: slender specimen, slip planes meet the free lateral surface. Right: squat specimen, slip planes meet the compression platens. (*b*) Rotation of slip planes towards an orientation normal to the load axis. Rotation is impeded at the platens. (*c*) Duplex slip, no lubrication at the platens. The slip planes are bent and the sample takes a barrel shape.

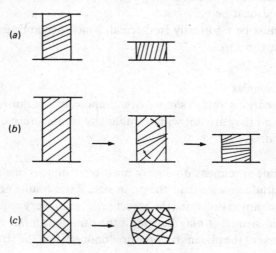

is avoided when there are two slip systems equally inclined to the load axis (duplex slip) but here, too, absence of lubrication results in barrelling of the sample for large strains and bending of the planes (figs. 1.11(c) and 1.12) (see also Bretheau & Dolin, 1978). For large strains the rotation of the planes must be taken into account when calculating the resolved shear stress (eq. (1.25)). Finally, it is unfortunately not superfluous to say that *the samples must be examined after deformation, at all scales of observation.*

1.2 Constitutive relations

1.2.1 State variables, time, strain

Constitutive relations describe the rheological properties of materials. We have already given (§1.1.2) characteristic constitutive equations for simple rheological behaviours, but most crystalline solids behave in a complex way and we have to decide which is the dominant rheology whose constitutive equation best describes the deformation. In particular, constant strain-rate tests produce stress–strain curves apparently independent of time and the theory of plasticity based on this kind of test is indeed a time-independent theory (e.g. Spencer, 1968). Creep tests, on the other hand, are interpreted in the framework of viscous flow. It must be emphasized that the difference is only apparent. Orowan (1940) was probably the first to point out that plastic properties of one material cannot be described by σ–ε curves (as is done in the theory of plasticity) but by a rate of flow $\dot{\varepsilon}$, at different stresses, temperatures and states of strain-hardening that depend not only on strain but on the whole history of the sample. Hart (1976), again, remarks that one must look for constitutive laws that can describe the time and temperature dependence of plastic flow and that *'the deformation that is commonly described as time-independent plasticity is in fact a rate process that is not qualitatively different from the high-temperature creep'.*

The most desirable general constitutive equation will therefore have the form:

$$f[\sigma, \dot{\varepsilon}, y(\varepsilon, \dot{\varepsilon} \ldots), T, P] = 0 \qquad (1.30)$$

where $\sigma, \dot{\varepsilon}, T, P$ are external state variables describing the state of the system and y is an internal state variable possibly depending on ε and its derivatives with respect to time $\dot{\varepsilon}, \ddot{\varepsilon}$, etc. (i.e. integrating all the history of the sample). Time itself must not explicitly appear in the constitutive equation if it is to be a true mechanical equation of state (describing unambiguously a state, as $PV = nRT$ does for perfect gases); the reason is that it is not a state variable but a path or coordinate variable, as a function of which the state

Fig. 1.12. Bending of planes in compressed samples of sodium chloride. Cleavage follows the bent {100} planes (not slip planes) normal and parallel to the load axis. Height of sample: top – 8.4 mm; bottom – 5.6 mm (courtesy V. Pontikis).

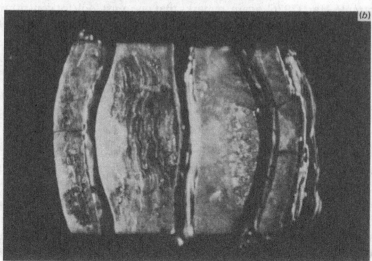

variables may vary; its status is not the same as that of T, σ or $\dot{\varepsilon}$: it is impossible to fix a value of time, as one fixes the temperature or the stress. Besides, *time and variables depending on time cannot simultaneously appear as independent variables and be treated as such in a constitutive relation* (McCartney, 1976).

The total plastic strain ε is not a good state variable and the state of strain-hardening cannot be described in terms of ε only. This is almost obvious if one realizes that the same total strain may result from very different histories which may have left completely different imprints on the internal state of the crystal.

1.2.2 Creep regimes: primary, quasi-steady-state

The general constitutive equation (1.30) is often put in the form:

$$\dot{\varepsilon} = \dot{\varepsilon}(\sigma, y, T, P) \tag{1.31}$$

The strain-rate $\dot{\varepsilon}$ is determined as the slope of the creep curve $\varepsilon(t)$; it is therefore necessary to examine the shape of the creep curve and see whether its slope can be constant (independent of time).

In compression tests, the creep curve normally starts by a *primary, transient* regime during which the creep-rate decreases in time; it is usually (but not always) followed by a quasi-steady-state regime during which the creep curve can be assimilated to a straight line within experimental scatter; eventually, as strain increases and the end constraints at the platens become important, the strain-rate decreases more markedly (fig. 1.13(*a*)). In tension, after the primary regime, the quasi-steady-state regime is followed by an unstable regime during which the strain-rate increases, the specimen suffers necking and eventually fails (fig. 1.13(*b*)). The primary regime was first described by Andrade (1910); he found that it could be fitted by the expression:

$$l = l_0 (1 + \beta t^{1/3}) \exp kt \tag{1.32}$$

where l_0 and l are respectively the initial length of the sample and its length at time t, β and k are constants. The equation of the creep curve, in its modern form $\varepsilon(t)$, is easily derived:

$$\varepsilon = \int_0^t \frac{dl}{l} = \varepsilon_0 + \ln(1 + \beta t^{1/3}) + kt \tag{1.33}$$

For small values of t, hence of ε, this is equivalent to:

▶ $$\varepsilon \approx \varepsilon_0 + \beta t^{1/3} + kt \tag{1.34}$$

The total strain is seen as the superposition of a primary transient creep in $t^{1/3}$ on a viscous steady-state creep with creep rate k.

The law in $t^{1/3}$, known as *Andrade's law*, gives a very good fit to the primary regime of practically all materials: steel, concrete, nylon, graphite, etc. (see Kennedy, 1962). It is clearly useless to try to interpret it in terms of a particular physical process.

The quasi-steady-state regime is usually treated as a steady-state regime and its creep-rate variation with stress and temperature is often taken as the equation of state of creep. Some caution, however, is in order: first, the fact that a portion of the creep curve appears linear or not over a certain time interval often depends on the time scale of the creep curve. It is apparent in fig. 1.14 that a portion of curve seen as concave towards the time axis on a certain scale becomes convincingly straight on a scale six times greater, and this without even taking the experimental scatter into account. The only way to decide whether it is appropriate to consider a quasi-steady-state regime is to plot the strain-rate against the true strain. If the strain-rate

Fig. 1.13. Typical creep curves. (*a*) Compression: the creep-rate decreases in time but a quasi-steady-state portion (II) can be found after the primary creep regime (I) and before the tertiary creep regime (III). (*b*) Tension: the creep-rate first decreases in the primary creep regime (I), a quasi-steady-state regime follows (II). The accelerating tertiary creep (III) leads to failure.

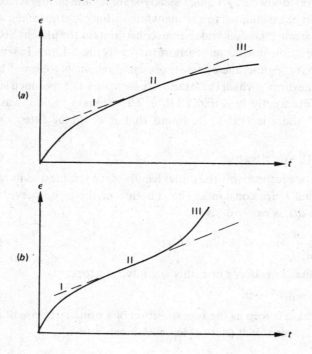

tends towards a limit for large strains, we can take this limit as quasi-steady-state creep-rate (fig. 1.15), it is clearly smaller than the apparently constant creep-rate of the final portion of the curve. Second, we are not sure that the minimum creep-rate thus determined is really a steady-state creep-rate and this is why we always use the prefix *quasi*. For instance, in the case of tensile tests, the minimum creep-rate in the region of the inflexion point can hardly be seen as a steady-state creep-rate since it increases eventually until failure occurs: a steady-state flow should be stable and never end. We will see in the next subsection that, in general, minimum creep-rate is not equivalent to steady-state creep-rate (if a steady state exists at all).

1.2.3 A mechanical equation of state

We shall, here, summarize the analysis of Hart (1970), which applies to the isothermal deformation of 'well-behaved' polycrystals (e.g. face-centred cubic metals like aluminium) deforming by intragranular slip. Hart investigates the conditions for a plastic equation of state to exist – it may be noted that its existence is completely independent of whether or not there exists a steady state.

Let us consider a relation between σ, ε and $\dot{\varepsilon}$ and describe incrementally the deformation history by:

▶ $$\mathrm{d} \ln \sigma = \gamma \, \mathrm{d}\varepsilon + m \, \mathrm{d} \ln \dot{\varepsilon} \tag{1.35}$$

where $\mathrm{d}\sigma$, $\mathrm{d}\varepsilon$, $\mathrm{d}\dot{\varepsilon}$ are the increments of stress, strain and strain-rate respectively. The logarithms are introduced for convenience reasons as we will use $\log \sigma$–$\log \dot{\varepsilon}$ plots.

Fig. 1.14. Steady-state regime? The same creep curve is shown using two different time scales. The decelerating creep curve A exhibits a spurious steady-state portion between t_1 and t_2 if the time scale is multiplied by 6 (curve B).

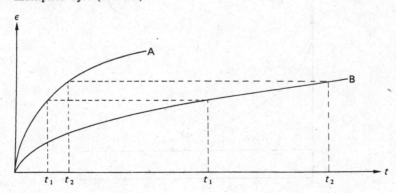

For a given path:

$$\gamma = \frac{\partial \ln \sigma}{\partial \varepsilon}\bigg)_{\dot{\varepsilon}} = \frac{\theta}{\sigma}\bigg)_{\dot{\varepsilon}} \tag{1.36}$$

$$m = \frac{\partial \ln \sigma}{\partial \ln \dot{\varepsilon}}\bigg)_{\varepsilon} \tag{1.37}$$

$\theta = \partial\sigma/\partial\varepsilon)_{\dot{\varepsilon}}$ is the work-hardening coefficient of the stress–strain curves, m is the strain-rate sensitivity of the stress.

Eq. (1.34) can be written:

$$d\varepsilon = \frac{1}{\gamma}\, d \ln \sigma - \frac{m}{\gamma}\, d \ln \dot{\varepsilon} \tag{1.38}$$

An equation of state exists if (1.38) can be integrated and if its solutions do not depend on the integration path. The condition is that γ and m be functions of σ and $\dot{\varepsilon}$ only:

Fig. 1.15. Quasi-steady-state regime can be identified on a creep curve $\varepsilon(t)$ by plotting the creep-rate against the creep strain. If $\dot{\varepsilon}(t)$ tends towards a limit, the limit can be taken as quasi-steady-state creep-rate $\dot{\varepsilon}_{QSS}$.

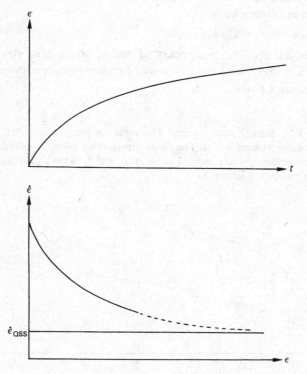

$$\left.\begin{array}{l} \gamma = \gamma(\sigma, \dot{\varepsilon}) \\ m = m(\sigma, \dot{\varepsilon}) \end{array}\right\} \tag{1.39}$$

It can be shown that if the condition (1.39) is fulfilled and, even if the more restrictive condition that $d\varepsilon$ be a total differential is not fulfilled, it is always possible to find an integrating factor F such that $F\,d\varepsilon$ is integrable. If $F \neq 1$, ε is not a good state variable.

Putting $dy = F\,d\varepsilon$ we can find a family of curves

$$y = y(\sigma, \dot{\varepsilon}) \tag{1.40}$$

solutions of the equation, that we will call *hardness curves*; the hardness y is the structural parameter introduced in the equation of state (1.31). In the log σ–log $\dot{\varepsilon}$ plot, constant-structure curves corresponding to given values of y can be drawn (fig. 1.16). The parameter y can be operationally defined in several ways all corresponding to the same state of hardening of the crystal. The meaning of the constant-structure lines is the following: if a sample is loaded at stress σ_0 and deforms at the rate $\dot{\varepsilon}_0$ and if, without changing the structure (i.e. with negligible strain), the stress is suddenly increased to σ_1 (or decreased to σ_2), then $\dot{\varepsilon}_0$ will change to $\dot{\varepsilon}_1$ or $\dot{\varepsilon}_2$ on the same constant-structure line. Stress relaxation allows a sampling of a wide range of stresses and strain-rates with very small plastic strain and is a choice technique (if well applied) to obtain these curves (Hart & Solomon, 1973). If the stress is changed by a stress jump during a creep test, usually by a small amount, one

Fig. 1.16. Typical constant-structure curves for various values of the hardness parameter y. Stress jumps and stress relaxation correspond to moving the representative point (σ, ε) on the same constant-structure curve.

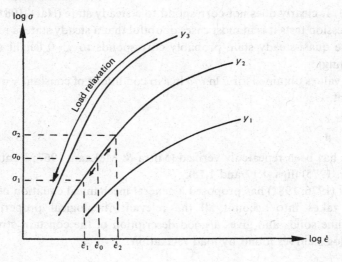

can obtain the slope of the curve (assimilated to a straight line), i.e. the strain-rate sensitivity of the stress:

$$m = \frac{\partial \ln \sigma}{\partial \ln \dot{\varepsilon}}\bigg)_y = \frac{\partial \ln \sigma}{\partial \ln \dot{\varepsilon}}\bigg)_\varepsilon \qquad (1.41)$$

This is a technique commonly used (see chapter 3).

There is, however, another way to plot log σ against log $\dot{\varepsilon}$ which leads to different curves and different values of the strain-rate sensitivity of the stress: in a constant-stress creep test, one can usually associate a value of the quasi-steady-state creep-rate with the value of the stress and plot log σ against log $\dot{\varepsilon}$ for different tests (i.e. different σ, $\dot{\varepsilon}$ couples). The curve obtained is not a constant-structure curve, since there is no reason for structure to remain constant along the creep curve as strain increases. If there exists a real steady state, then it must correspond to zero strain-hardening in constant strain-rate tests, hence $\gamma = 0$ in (1.35). The curve obtained in the log σ-log $\dot{\varepsilon}$ plot is the $\gamma = 0$ curve, different from the $y = $ const. curves (fig. 1.16). Its slope $1/n$ is the strain-rate sensitivity of the stress in steady state but it is different from m. (We have called the slope $1/n$ because it is common to plot log $\dot{\varepsilon}$ against log σ and obtain n as the stress sensitivity of the strain-rate. It is the stress exponent of the power law $\dot{\varepsilon} = A\sigma^n$.)

In tension tests the minimum creep-rate corresponds to the start of the plastic instability at the inflexion point of the creep curve, for which we have the Considère criterion (see below):

$$\frac{\partial \sigma}{\partial \varepsilon}\bigg)_{\dot{\varepsilon}} = 0 \qquad (1.42)$$

or $\gamma = 1$. It clearly does not correspond to a steady state (Hart, 1981). For compression tests it is in most cases doubtful that a steady state is reached and the quasi-steady state probably corresponds to $\gamma > 0$ (slight strain-hardening).

The values obtained for $\partial \ln \sigma / \partial \ln \dot{\varepsilon}$ in conditions of constant y or γ are different. Usually,

$$\frac{1}{n} > m \qquad (1.43)$$

This has been repeatedly verified (Mitra & McLean, 1967; Pontikis & Poirier, 1975) (figs. 1.17 and 1.18).

Hart (1976, 1981) has proposed a general mechanical equation of state, which takes into account all the relevant rheological properties of crystalline solids and gives a good description of the constant structure log σ–log $\dot{\varepsilon}$ curves found by load relaxation:

$$\ln\left(\frac{\sigma^*}{\sigma}\right)=\left(\frac{\dot{\varepsilon}^*}{\dot{\varepsilon}}\right)^{\lambda} \tag{1.44}$$

with

$$\dot{\varepsilon}^*=\left(\frac{\sigma^*}{\mu}\right)^{n}f\exp\left(-\frac{Q}{RT}\right) \tag{1.45}$$

σ^* is the hardness (it is our y, defined here in an appropriate way), λ is a material constant ($\lambda=0.15$ for aluminium), f is a rate constant, Q describing the temperature dependence is an activation energy (see chapter 3), μ is the shear modulus and n is the stress sensitivity of quasi-steady-state creep-rate.

The Hart equation of state also accounts for the experimental fact that there is a scaling relation with respect to σ and $\dot{\varepsilon}$ among the curves of different hardness: the constant-hardness curves can all be brought into coincidence by translations along straight lines of slope m (Hart & Solomon, 1973).

1.3   Stability of deformation

Plastic instability occurs when it becomes easier to deform the sample where it has already been deformed than elsewhere; conversely the deformation is stable, when it is easier to start deformation elsewhere than pursue it where it has begun. Plastic instability, therefore, always manifests itself by a localization of deformation, whereas stable deformation must be uniform at the scale of the sample.

Fig. 1.17. Constant strain-hardening curves ($\gamma=$const.) including the steady-state curve ($\gamma=0$) do not give the same value of the strain-rate sensitivity as the constant-structure curves ($y=$const.).

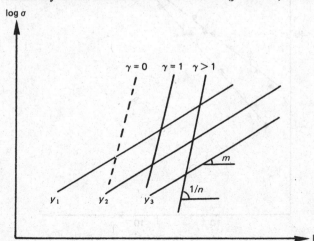

The appearance of plastic instability and the conditions leading to it have been the object of detailed analyses (see Hart, 1967; Jonas *et al.*, 1976). We will only derive a simple criterion for instability in tension and shear and review the physical conditions leading to it.

Let us consider, for example, the case of constant strain-rate traction tests (the conclusions are identical for creep tests); for an increment of strain along the stress–strain curve, we have from eq. (1.35):

$$\text{d} \ln \sigma = \gamma \, \text{d}\varepsilon \tag{1.46}$$

since d ln $\dot{\varepsilon} = 0$; γ is still given by (1.36).

Now if A is the cross-section area of the sample, and F the force applied to the sample, we have:

$$F = \sigma A \tag{1.47}$$

We will choose for an instability criterion the condition that the load-carrying capacity decrease with increasing strain, i.e.:

Fig. 1.18. Minimum creep-rate curve (A) and constant-structure curve (B) for AgCl single crystals (after Pontikis & Poirier, 1975). Slope of A – $1/m = 1/3$; slope of B – $1/n = 1/5.3$.

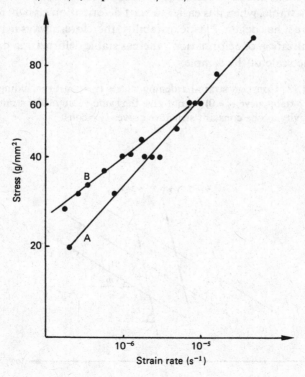

$$\frac{d \ln F}{d\varepsilon} \leqslant 0 \tag{1.48}$$

From (1.47), we have:

$$\frac{d \ln F}{d\varepsilon} = \frac{d \ln \sigma}{d\varepsilon} + \frac{d \ln A}{d\varepsilon} \tag{1.49}$$

The instability may appear for d ln $F/d\varepsilon = 0$, i.e.:

$$\frac{d \ln \sigma}{d\varepsilon} = -\frac{d \ln A}{d\varepsilon}$$

Plastic deformation occurs at constant volume, so for a sample of length l and cross-section area A:

$$Al = \text{const}$$

or

$$d \ln A + d \ln l = d \ln A + \frac{dl}{l} = d \ln A + d\varepsilon = 0$$

or

$$\frac{d \ln A}{d\varepsilon} = -1$$

hence:

$$\frac{d \ln \sigma}{d\varepsilon} = \gamma = 1 \tag{1.50}$$

or:

▶ $$\frac{d\sigma}{d\varepsilon} = \sigma \tag{1.51}$$

This is known as the Considère criterion; it defines the stress above which necking occurs in a stress–strain test (fig. 1.19).

In shear, where the area A remains constant, we would have:

$$\frac{d \ln \sigma}{d\varepsilon} = 0 \tag{1.52}$$

The variation of σ (or log σ) with strain can be decomposed into two kinds of terms: the terms that express strain-hardening and those that express strain-softening; they correspond to various physical processes. *Instability will occur when the sum of the strain-softening terms (negative) balances the sum of the strain-hardening terms (positive), so that there is a net strain-softening.* Obviously it is then easier to deform an already strained region (i.e. at lower stress) than to start deformation elsewhere. There are several potential strain-softening processes (Jonas & Luton, 1978; Poirier, 1980) of which the most important are:

– Dynamic recrystallization, a process whereby strain-hardened grains are replaced by new unstrained grains (see chapter 6). This leads to oscillations, or peaks, in the stress–strain curve, hence to strain–softening in the regions with negative slope (Rossard & Blain, 1958). It has been shown, in the case of magnesium, that

Fig. 1.19. Considère construction. The strain for which necking can occur is determined by sliding a segment of unit length *AB* along the axis of abscissae, until a tangent drawn from *A* touches the curve at *C*, above *B* ($d\sigma/d\varepsilon = \tan \alpha = CB/AB = \sigma/1$).

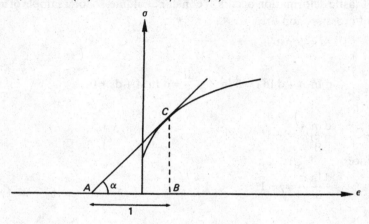

Fig. 1.20. Strain softening due to dynamic recrystallization in magnesium causes plastic instability (shear bands) (after Ion *et al.*, 1982).

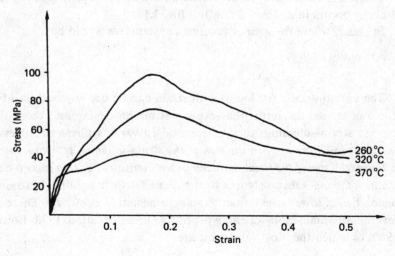

under conditions leading to a stress peak ($T = 260°C$, $\dot{\varepsilon} = 10^{-5}/s$) the dynamically recrystallized grains were concentrated in a single shear band which accommodated all subsequent deformation (Ion et al., 1982) (fig. 1.20).

– Adiabatic heating, when deformation is thermally activated (see chapter 3) if there is little or no heat exchange during deformation, owing to low thermal diffusivity and/or rapid deformation, the heat dissipated during straining may raise the temperature of the sample enough to make deformation considerably easier.

Readings

W. D. Means, *Stress and strain*, Springer-Verlag, New York (1976), 339 pp. A crystal-clear introduction to basic concepts of continuum mechanics for geologists.

M. Reiner, *Deformation, strain and flow*, H. K. Lewis, London (1969), 347 pp. An illuminating introduction to rheology by one of its founding fathers.

A. H. Cottrell, *The mechanical properties of matter*, Wiley, New York (1964), 430 pp. From physics to mechanics of materials at the undergraduate level. A gem of a book.

2

The agents of deformation: lattice defects

Plastic deformation of crystals is carried by lattice defects: the migration of vacancies is responsible for deformation by transport of matter, the propagation of dislocations is responsible for deformation by slip, in polycrystals grain-boundary motion may also be a source of deformation.

Vacancies are point defects: vacant sites in the crystal lattice. For entropic reasons there exists an equilibrium concentration of vacancies dependent on the temperature. The vacancies migrate by exchange with neighbouring ions; their migration obeys diffusion equations (Fick's laws) identical to those ruling the diffusion of heat.

Dislocations are linear defects at the boundary between an area on which slip has taken place and the rest of the crystal, as yet 'unslipped'. They create an internal strain field and stress field extending through the whole crystal and whose strength decreases as the inverse of the distance to the dislocation. It is through their strain field that dislocations 'see' an applied stress and move, thus increasing the slipped area. Dislocation motion is impeded by a thermally activated lattice friction force (Peierls' force) and by obstacles. Orowan's equation is a microscopic equation of state relating the strain-rate to the dislocation density velocity.

Grain boundaries in a chemically homogeneous material are two-dimensional defects separating grains whose lattices have different orientations. They can often be described as arrays of dislocations. Recrystallization is a change in the granular structure involving motion of grain boundaries.

2.1   Generalities

A perfect crystal, i.e. a crystal whose lattice periodicity would be nowhere disturbed, will respond elastically to an applied shear stress and will acquire no permanent strain as long as the stress remains inferior to the theoretical elastic limit, high enough to break all chemical bonds simultaneously across a slip plane and allow plastic slip as described in § 1.1.2.

The theoretical elastic limit is of the order of a tenth of the shear modulus ($\mu/10$) for all crystals; under these conditions, deformation would involve such high stresses that it would become practically impossible: there would be no metallurgy and no ductile tectonics in a world of perfect crystals. It is clearly not the case and the explanation for the relative ease of deformation in crystals lies in the fact that crystals do not deform all at once: strain occurs progressively – a view consistent with the realization that plasticity is a flow process – at a lesser cost in power. Real crystals are not perfect but contain defects where periodicity is locally broken; some of the most important defects are potentially deformation 'carriers': their motion in a stress field produces an elementary deformation (a quantum of strain, so to speak) in the sense that makes the applied forces do mechanical work. In principle, if we know the values of the quantum of strain corresponding to a given defect (i.e. its strength), the concentration of defects and their velocity, we can obtain a microscopic equation of state for steady state (similar to Ohm's law for electrical current): flow rate = concentration of carriers × strength × velocity.

(i) Vacancies are point defects – vacant sites in the crystal lattice; they can migrate by exchanging ions with neighbouring sites. We will see that in a given stress regime, there exist gradients in the vacancy concentration; vacancies flow down the concentration gradient, which corresponds to a flow of matter in the opposite sense, achieving deformation of the crystals by what is known as diffusion creep (see chapter 7). Vacancies are the agents of deformation by transport of matter.

(ii) Dislocations are line defects bounding areas in the crystal where slip has already taken place. The outward propagation of dislocations under stress makes the 'slipped' area increase at the expense of the as-yet unslipped crystal. Dislocations are the agents of deformation by slip.

(iii) Grain boundaries are two-dimensional defects separating crystals of different lattice orientations. In certain conditions, their shear strength can be relaxed and adjoining grains can slide along the grain boundary, which results in macroscopic straining with little deformation inside the grains (much in the way of a wet sandbag). Grain boundaries are the agents of intergranular deformation.

In the following sections, we will give a brief and elementary account of these three types of defects, introducing at the arm-waving level the features or properties that will be used later, and leaving aside rigorous general treatments and mathematical developments, which the reader will find in articles (given in references) and in specialized books (see list of readings).

2.2   Vacancies and diffusion

2.2.1   Thermodynamics

A vacancy is a vacant lattice site in a crystal. To define the thermodynamic quantities that we will use later, let us 'prepare' a crystal containing one vacancy by the following thought experiment (fig. 2.1):

(i) Take a piece of crystal containing n atoms at P and T (for simplicity we will consider a chemical element with only one type of atoms).

(ii) Extract an atom from inside by cutting the Z bonds linking it to its neighbours (Z is the coordination), thus leaving a vacancy.

(iii) Deposit the atom on the surface, at a step on a ledge, so that the external area is left unchanged; some bonds are restored in the process. The total number of atoms has not changed but the process of formation of one vacancy is accompanied by a variation of volume V, internal energy E and entropy S (between final and initial state).

The variation of volume is obviously close to the value of the atomic volume Ω, but it is not equal to it owing to a relaxation of the crystal around the vacancy (inward or outward); it is called the *volume of formation* of a vacancy, ΔV_f. The variation of energy corresponds to the energy of the net numbers of broken bonds; it is called *energy of formation* of a vacancy, ΔE_f. Finally, the *entropy of formation*, ΔS_f, corresponds to changes in the vibrational atomic frequencies in the vicinity of the defect.

We can therefore define the *enthalpy*, ΔH_f, and the Gibbs *free energy*, ΔG_f, *of formation* of one vacancy:

▶ $$\Delta H_f = \Delta E_f + P \Delta V_f \tag{2.1}$$

▶ $$\Delta G_f = \Delta H_f - T \Delta S_f \tag{2.2}$$

Fig. 2.1. Formation of one vacancy (principle): An atom inside the crystal is brought to a step on a ledge at the surface (the volume increases but not the area).

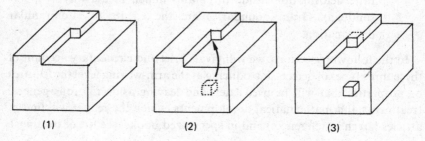

(1)   (2)   (3)

If n_v vacancies are formed, the change in free energy of the crystal (with the perfect crystal as reference state) is:

$$\Delta G = n_v \, \Delta G_f \tag{2.3}$$

but to write the free energy change ΔG, we must take into account a configurational entropy term ΔS_c due to the fact that the n_v vacancies can be distributed among the n sites in W different ways:

$$\Delta G = n_v \, \Delta G_f - T \Delta S_c \tag{2.4}$$

Now by definition:

$$\Delta S_c = k \ln W \tag{2.5}$$

where k is Boltzmann's constant. As vacancies are indistinguishable, W is equal to the number of ways n_v objects can be selected from a set of n identical objects:

$$W = \frac{n(n-1)(n-2)\cdots(n-n_v+1)}{1\cdot 2\cdot 3\cdot \ldots \cdot n_v} = \frac{n!}{(n-n_v)! \, n_v!} \tag{2.6}$$

A good approximation of ΔS_c can be obtained using Stirling's formula:

$$\ln(x!) \approx x \ln x - x$$

and by making the reasonable assumption that $n_v \ll n$; eq. (2.4) finally can be written:

$$\Delta G \approx n_v \, \Delta G_f - kTn_v \left(1 + \ln \frac{n}{n_v}\right) \tag{2.7}$$

We see that as n_v increases, the increase in entropy can balance the increase in enthalpy so that, for a given temperature, there is a value of n_v that corresponds to a minimum in free energy, i.e. to an equilibrium situation, for $\partial \Delta G/\partial n_v = 0$.

The value of the equilibrium atomic fraction of vacancies for a given temperature is:

▶ $$N_v \equiv \frac{n_v}{n} = \exp\left(-\frac{\Delta G_f}{kT}\right) \tag{2.8}$$

2.2.2   Diffusion

Diffusion of vacancies by exchange with atoms is a random-walk process, described by Einstein's relation:

$$\langle R^2 \rangle = \Gamma t \delta l^2 \tag{2.9}$$

where $\langle R^2 \rangle$ is the mean square distance covered by a vacancy during time t, Γ is the jump frequency and δl the jump distance. The *diffusion coefficient of vacancies* is defined by:

$$D_v = \frac{\langle R^2 \rangle}{6t} = \frac{\Gamma}{6} \delta l^2 \tag{2.10}$$

The jump frequency (total number of jumps of a vacancy per second) is a thermally activated quantity proportional to an attempt frequency v ($\approx 10^{13}$ s) and to the probability that an atom, assisted by thermal energy, succeeds in overcoming the potential barrier:

$$\Gamma = Zv \exp\left(-\frac{\Delta G_m}{kT}\right) \tag{2.11}$$

where Z is the coordination and ΔG_m the *free energy of migration*. The jump distance depends on the crystal structure and the direction. For cubic crystals of cell parameter a, it is easily shown that: $Z\delta l^2/6 = a^2$. In general, D_v is written:

▶ $$D_v = D_{0v} \exp\left(-\frac{\Delta H_m}{kT}\right) \tag{2.12}$$

where ΔH_m is the activation enthalpy for migration and D_{0v} includes the entropic term. The self-diffusion coefficient is the diffusion coefficient of the atoms of the crystal (in the case of compounds one can define a self-diffusion coefficient for each species). *The self-diffusion coefficient is not equal to the diffusion coefficient of the vacancies*: a vacancy is surrounded by atoms and its probability to jump is equal to the number of neighbouring atoms times the probability for any of the neighbours to jump into the vacancy; an atom in the crystal can jump only if a vacancy is on a nearest-neighbour site and the probability of that occurrence is equal to the atomic fraction of vacancies present in the crystal. The self-diffusion coefficient D_{sd} can be written:

▶ $$D_{sd} = D_v N_v \tag{2.13}$$

or

$$D_{sd} = D_0 \exp\left(-\frac{\Delta H_f + \Delta H_m}{kT}\right) \tag{2.14}$$

The quantity

$$\Delta H_{sd} = \Delta H_f + \Delta H_m \tag{2.15}$$

is the *activation enthalpy for self-diffusion*; it can be written

$$\Delta H_{sd} = \Delta E_{sd} + P\Delta V_{sd} \tag{2.16}$$

where ΔV_{sd} is the *activation volume for self-diffusion* (see chapter 4). ΔH_{sd} and ΔV_{sd} can be experimentally determined by measuring the diffusion coefficients for different temperatures and pressures (fig. 2.2)

$$\Delta H_{sd} = -k\frac{\partial \ln D_{sd}}{\partial\left(\dfrac{1}{T}\right)} \tag{2.17}$$

$$\Delta V_{sd} = -kT\frac{\partial \ln D_{sd}}{\partial P} \tag{2.18}$$

Values of D_0 and ΔH_{sd} for typical metals, ceramics and minerals are given in table 2.1. See also Freer (1981) for a review on diffusion in silicate minerals.

The mathematics of diffusion being universal, whatever the particle or physical quantity that diffuses, vacancy diffusion obeys the two laws (equivalent to Fourier's laws in the case of heat conduction) called here Fick's laws:

▶ $$J = -D_v \,\text{grad}\, C \tag{2.19}$$

▶ $$\frac{\partial C}{\partial t} = D_v \nabla^2 C \tag{2.20}$$

or in the simple case of unidirectional diffusion (along axis Ox)

$$J = -D_v \frac{\partial C}{\partial x} \tag{2.21}$$

$$\frac{\partial C}{\partial t} = D_v \frac{\partial^2 C}{\partial x^2} \tag{2.22}$$

C is the concentration of vacancies (number per unit volume) related to the atomic fraction by:

Fig. 2.2. Determination of the activation enthalpy and activation volume for diffusion. (a) Activation enthalpy ΔH_{sd} is determined from the slope of the plot $\ln D$ against $1/T$ at constant pressure.
(b) Activation volume ΔV_{sd} is determined from the slope of the plot $\ln D$ against P at constant temperature.

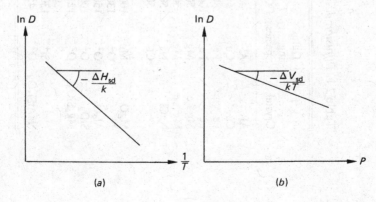

        (a)   (b)

Table 2.1. *Diffusion data for a few typical crystals*

Crystal	Diffusing species	T (°C)	D_0 (cm²/s)	ΔH (kcal/mol)	$D(T_{max})$ (cm²/s)	Reference
Al	Al	460–650	1.71	34	1.71×10^{-8}	Lundy & Murdock, J. Appl. Phys., **33**, 1671 (1962)
Cu	Cu	685–1062	0.20	47.1	4.36×10^{-9}	Kuper et al., Phys. Rev, **98**, 1870 (1961)
Fe	Fe	809–905	1.9	57.2	5.43×10^{-11}	Buffington et al., Acta Metall., **9**, 434 (1961)
Fe	Fe	1064–1393	0.18	64.5	7.05×10^{-10}	Buffington et al., Acta Metall., **9**, 434 (1961)
Mg ∥ c	Mg	500–630	1.78	33.2	1.66×10^{-8}	Combronde & Brebec, Acta Metall., **19**, 1393 (1971)
Mg ⊥ c	Mg	500–630	1.75	33	1.83×10^{-8}	Combronde & Brebec, Acta Metall., **19**, 1393 (1971)
NaCl	Na⁺	500–750	33.16	45.5	7.15×10^{-9}	Bénière et al., J. Phys. Chem. Sol., **31**, 825 (1970)
	Cl⁻	500–750	61	49.2	2.18×10^{-9}	Bénière & Chemla, C.R. Acad. Sci. Paris, **267**, 633 (1968)
MgO	Mg⁺	1000–2400	4.19×10^{-4}	63.5	2.92×10^{-9}	Wuensch et al., J. Chem. Phys., **58**, 5258 (1973)
	O²⁻	1300–1750	2.5×10^{-6}	62.4	5.01×10^{-13}	Oishi & Kingery, J. Chem. Phys., **33**, 905 (1960)
Al₂O₃	O²⁻	1500–1780	1900	152	1.59×10^{-13}	Oishi & Kingery, J. Chem. Phys., **33**, 480 (1960)
SiO₂	O²⁻	1000–1220	3.7×10^{-9}	55	3.7×10^{-17}	Haul & Dumbgen, Z. Electro. Chemie, **66**, 636 (1962)
Mg₂SiO₄	O²⁻	1275–1625	3.5×10^{-3}	89	29.3×10^{-14}	Reddy et al., J. Geophys. Res., **85**, 322 (1980)
		1300–1600	2.3×10^{-6}	70		Jaoul et al., J. Geophys. Res., **88**, 613 (1983)
	Si⁴⁺	1300–1700	1.5×10^{-6}	90	1.87×10^{-16}	Jaoul et al., 'Anelasticity in the Earth' (1981)
NaAlSi₃O₈	O²⁻	450–750	9.8×10^{-6}	33.4	7.97×10^{-13}	Yund et al., Phys. Chem. Min., **7**, 185 (1981)

$$C = \frac{N_v}{\Omega} \tag{2.23}$$

where Ω is the atomic volume.

J is the flux of vacancies (number per second flowing through unit area) flowing down the concentration gradient, which represents here the driving force.

2.2.3 Diffusion in binary ionic crystals

In ceramics and minerals the problem of diffusion of the species constituting the crystal is considerably more complicated than in metals. It is always possible to describe the diffusion of one given species by exchange with vacancies on its own sublattice, but there are additional constraints arising from the fact that the crystal is constituted by ions with electrical charges that are not screened by a sea of almost free electrons as in metals. As a consequence, extraction of an ion leaves an oppositely charged vacancy. However, (i) the crystal must remain electrically neutral, which imposes a relation between the concentrations of the vacancies of anions and cations, and (ii) there must be no net electrical current flowing, which imposes a relation between the fluxes of anions and cations.

In addition to these conditions stemming from the existence of charged ions, there is another complication of thermodynamic nature in the general case when more than two species diffuse in the same mineral: the flux of one ion depends on the concentration gradients of all others; the diffusion can be described in the framework of irreversible thermodynamics by a system of linear equations relating the fluxes of species i to the chemical potential gradient of species j and diffusion coefficient matrices must be introduced. (See Lasaga, 1979, for the case of multicomponent diffusion in silicates.)

We will restrict ourselves here to the much simpler case of binary ionic crystals (alkali halides and oxides) and establish some of the results that we will use later on.

(i) *Vacancy concentration and diffusion regimes*

Let us consider a pure binary ionic compound, whose anions and cations have equal and opposite charges (e.g. Na^+Cl^- or $Mg^{2+}O^{2-}$). If diffusion occurs by exchange with vacancies, the diffusion coefficients of cations and anions on their own sublattices will be given by relations identical to (2.13):

$$\begin{aligned} D_c &= D_{vc} N_c \\ D_a &= D_{va} N_a \end{aligned} \tag{2.24}$$

where D_c and D_a, N_c and N_a, D_{vc} and D_{va} are the coefficients of self-diffusion of cations and anions, and the atomic fractions and diffusion coefficients of cationic and anionic vacancies respectively. Thermal cationic and anionic vacancies are formed in pairs (Schottky pairs) and it can be shown that at equilibrium the product of their atomic fractions is a constant depending only on temperature (the 'solubility product' of the pair, so to speak):

$$N_a N_c = N_0^2 = \exp\left(\frac{\Delta S_f}{kT}\right) \exp\left(-\frac{\Delta H_f}{kT}\right) \tag{2.25}$$

where N_a, N_c are the atomic fractions of anionic and cationic vacancies, and ΔS_f and ΔH_f the entropy and enthalpy of formation of the Schottky pairs.

With the electrical neutrality condition

$$N_a = N_c \tag{2.26}$$

we find:

$$N_a = N_c = \exp\left(\frac{\Delta S_f}{2kT}\right) \exp\left(-\frac{\Delta H_f}{2kT}\right) \tag{2.27}$$

The activation enthalpies for diffusion of the anions and cations are respectively, after (2.15):

$$\Delta H_a = \frac{\Delta H_f}{2} + \Delta H_{ma} \tag{2.28}$$

$$\Delta H_c = \frac{\Delta H_f}{2} + \Delta H_{mc} \tag{2.29}$$

ΔH_{ma} and ΔH_{mc} are the migration enthalpies of anions and cations. The cations, smaller in general, are often more mobile than the anions; they give the principal contribution to the ionic electrical conductivity which in first approximation is proportional to the diffusion coefficient of the cations. This diffusion regime, where vacancies are thermally created, is the only one possible for theoretically pure crystals and even for slightly impure crystals at high temperature, as we will see presently. Let us now consider the same crystal but with a small concentration of aliovalent cationic impurities in substitution (for instance Ca^{2+} in NaCl). We consider only cationic impurities because anionic impurities are often too big to enter in substitution into the lattice, but the reasoning would be the same.

If a Ca^{2+} ion replaces a Na^+ ion, there is an excess of one positive charge, which is compensated by the non-thermal creation of one negatively charged cationic vacancy (*extrinsic vacancy*). If the atomic fraction of impurities is C, the electrical neutrality condition is written:

$$N_a + C = N_c \tag{2.30}$$

Hence, with (2.25):

$$N_0^2 = N_a(C + N_a) \tag{2.31}$$

The solutions of this equation are:

$$N_a = \frac{C}{2}\left[\left(1 + \frac{4N_0^2}{C^2}\right)^{1/2} - 1\right] \tag{2.32}$$

$$N_c = \frac{C}{2}\left[\left(1 + \frac{4N_0^2}{C^2}\right)^{1/2} + 1\right] \tag{2.33}$$

If $C \ll N_0$, i.e. if the atomic fraction of impurities is very small compared with that of thermal vacancies, diffusion is controlled by the formation and migration of thermal vacancies and the activation enthalpies are given by (2.28) and (2.29). This is the *intrinsic* regime.

If $C \gg N_0$, the concentration of thermal vacancies becomes negligible and the concentration of cationic vacancies is now fixed by the concentration of impurities:

$$N_c \approx C \tag{2.34}$$

The vacancies are freely available for diffusion, no formation enthalpy has to be spent and

$$\Delta H_c = \Delta H_{mc} \tag{2.35}$$

Conversely, the concentration of anionic vacancies is much reduced and the self-diffusion coefficient of the anions decreases. This is the *extrinsic* regime, which occurs for impure crystals below a temperature that depends on the impurity content (the more impure the crystal, the more the extrinsic regime extends to high temperatures). For still lower temperatures, or more impure crystals, the extrinsic vacancies may remain bound to the impurity ion whose extra charge they compensate. In that case, they become available for diffusion only if the complex is dissociated and we have:

$$\Delta H_c = \frac{\Delta H_b}{2} + \Delta H_{mc} \tag{2.36}$$

where ΔH_b is the binding energy of the impurity–vacancy complex.

The intrinsic, extrinsic and association regimes are responsible for domains with different slopes on the Arrhenius plot for cation diffusion (or ionic electrical conductivity) (fig. 2.3).

It is obviously unwise to extrapolate diffusion coefficients to higher or lower temperatures, when the whole Arrhenius plot is not known.

(ii) *Influence of oxygen partial pressure on diffusion in oxides*

The concentration of vacancies is often sensitive to oxygen partial pressure, in the case of oxides. As an example of one of the possible situations, let us examine the case of a binary oxide where the metallic ion can react with gaseous oxygen to increase its oxidation degree. In the case of wüstite FeO we have:

$$\tfrac{1}{2}O(g) + 2Fe^{2+} \rightarrow O^{2-} + V_{Fe}^{2'} + 2Fe^{3+} \tag{2.37}$$

$V_{Fe}^{2'}$ represents a vacancy in the Fe^{2+} sublattice, with effective charge $2-$; Fe^{3+} is a ferric ion in the Fe^{2+} sublattice associated with a hole (h).

The law of mass action for reaction (2.37) gives

$$\frac{[V_{Fe}^{2'}][h]^2}{p_{O_2}^{1/2}} = K \tag{2.38}$$

and charge conservation imposes

$$[V_{Fe}^{2'}] = [h] \tag{2.39}$$

hence

$$[V_{Fe}^{2'}] \propto p_{O_2}^{1/6}$$

If the cations Fe^{2+} diffuse by a vacancy mechanism, we see that their self-diffusion coefficient D_{Fe} depends on oxygen partial pressure:

$$D_{Fe} \propto [V_{Fe}^{2'}] \propto p_{O_2}^{1/6} \tag{2.40}$$

See Chen & Peterson (1980) for the more general case of magnesio-wüstite $(Mg_xFe_{1-x})O$.

(iii) *Transport of matter in binary ionic crystals*

Ruoff (1965) has treated the case of binary ionic crystals $A_\alpha B_\beta$, stoichiometric and where both species diffuse by a vacancy mechanism.

Fig. 2.3. Diffusion regimes in ionic crystals. I. Intrinsic regime (thermal vacancies) $\Delta H = \Delta H_f/2 + \Delta H_m$. II. Extrinsic regime (charge-compensating vacancies) $\Delta H = \Delta H_m$. III. Association regime (impurity-vacancy pairs) $\Delta H = \Delta H_b/2 + \Delta H_m$.

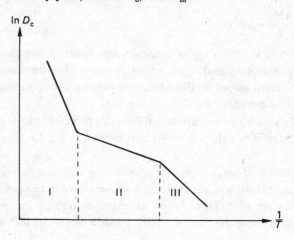

Transport of matter by diffusion implies that a molecular group must be transported, although not as a unit.

For steady state, the flux J_i of vacancies of the i species ($i = A, B$) is given by:

$$J_i = C_i v_i = C_i M_i F_i \qquad (2.41)$$

where C_i is the concentration of vacancies, v_i their drift velocity, which is equal to their mobility M_i times the driving force F_i for diffusion.

The driving force can be expressed in terms of the chemical potential μ_i of the vacancies, of their charge q_i and of the electric field E

$$F_i = -(\text{grad } \mu_i + q_i E) \qquad (2.42)$$

With Nernst relation relating the diffusion coefficient D to the mobility

$$D_i = M_i k T \qquad (2.43)$$

we have

$$J_i = -\frac{C_i D_i}{kT}(\text{grad } \mu_i + q_i E) \qquad (2.44)$$

The condition of zero net current is written:

$$q_A J_A + q_B J_B = 0 \qquad (2.45)$$

For intrinsic diffusion, we have:

$$\beta C_A = \alpha C_B \qquad (2.46)$$

and

$$\alpha q_A + \beta q_B = 0 \qquad (2.47)$$

Taking the expression of the chemical potential

$$\mu_i = \mu_i^0 + kT \ln\left(\frac{C_i}{C_i^0}\right) \qquad (2.48)$$

we have

$$\text{grad } \mu_i = kT \frac{\text{grad } C_i}{C_i} \qquad (2.49)$$

which, with eqs. (2.44) to (2.47) gives:

$$J_A = -(D_A \text{ grad } C_A)\frac{(\alpha + \beta)D_B}{\beta D_A + \alpha D_B} \qquad (2.50)$$

Using the atomic fractions of A and B $- n_A = \alpha/(\alpha + \beta)$ and $n_B = \beta/(\alpha + \beta)$ – we can define an *effective diffusion coefficient*:

$$D' = \frac{D_A D_B}{D_A n_B + D_B n_A} \qquad (2.51)$$

The fluxes of A and B can then be written, using Fick's law with the same effective diffusion coefficient D':

$$J_i = -D' \text{ grad } C_i \qquad (2.52)$$

2.2.4   Short-circuit diffusion

Vacancies are point defects and it is easily conceivable that they may migrate along the highly imperfect cores of linear or two-dimensional crystalline defects, more rapidly than in the bulk of the crystal. Indeed, there is good evidence that vacancies diffuse easily along dislocations (pipe diffusion) and grain boundaries (grain-boundary diffusion). Experimental measurements of short-circuit diffusivity are difficult and the data are scarce; it is, however, possible to reach some conclusions:

(i) Pipe diffusion

Self-diffusion coefficients in dislocations are many orders of magnitude larger than in the lattice. For instance, the pipe-diffusion coefficient of oxygen in albite is five orders of magnitude larger than in bulk (Yund et al., 1981). The quantity obtained by fitting experimental results to models is $D_d A_d$ (in cm^4/s) where D_d is the diffusion coefficient along the dislocation and A_d the effective cross-sectional area of the pipe (considering the dislocation as a tube). As for bulk diffusion, it is possible to write:

$$D_d A_d = D_{0d} A_d \exp\left(-\frac{\Delta H_d}{kT}\right) \tag{2.53}$$

The activation energy ΔH_d is lower than ΔH for bulk self-diffusion (see Balluffi, 1970). Typically: $0.4 < \Delta H_d/\Delta H < 0.7$, e.g.:

Ag $\Delta H_d = 17.0$ kcal/mol $\Delta H = 45$ kcal/mol

Al $\Delta H_d = 19.6$ kcal/mol $\Delta H = 30.1$ kcal/mol

Pipe diffusion will therefore be important at lower temperatures, when bulk diffusion is very slow.

(ii) Grain-boundary diffusion

Diffusivity along grain boundaries is also considerably faster than in the lattice. The experimental results are given in terms of $D_{gb}\delta$ where δ is an effective width of the grain boundary, which here, too, is a phenomenological parameter without much physical significance (see Martin & Perraillon, 1980).

$$D_{gb}\delta = D_{0gb}\delta \exp\left(-\frac{\Delta H_{gb}}{kT}\right) \tag{2.54}$$

Here, too, the activation energy is of the order of half the activation energy for lattice diffusion (table 2.2). Grain-boundary diffusion is a process dominant at lower temperatures. Diffusion may be considerably faster in migrating grain boundaries than in static ones (Smidoda et al., 1978).

2.3 Dislocations and slip

2.3.1 Definitions

We can operationally *define a dislocation loop as the line that bounds a closed area over which slip has taken place* (i.e. the part of the solid on one side of the area has been translated with respect to the other side by a slip vector *b*). As the loop expands under stress (see below) the slipped area increases and so does the shear strain. The dislocation loop is characterized by the slip vector or *Burgers vector b*, which is quantized in the case of a crystal (it can only be a vector of the Bravais lattice). A segment of dislocation normal to its Burgers vector has an *edge* character and the dislocation line lies at the edge of an extra half plane (fig. 2.4); a segment of dislocation parallel to its Burgers vector has a *screw* character and the atomic planes are distorted in the shape of a helicoid whose axis is the dislocation line (fig. 2.4). An edge segment can only slip on the slip (or glide) plane defined by the dislocation line and the Burgers vector and it slips normally to itself; a screw segment also slips normally to itself but can theoretically slip on any atomic plane containing it. When the loop reaches the surface of the crystal it leaves a step of width *b* (fig. 2.5).

A dislocation is a linear defect, it distorts the atomic planes in the crystal (see fig. 2.4), and therefore introduces an internal strain.[†] (The internal strain is caused by the presence of a dislocation; it is 'exteriorized', so to speak, as the dislocation leaves the crystal.) A dislocation is therefore the source of an internal stress field, related to the strain field by Hooke's law. It can be shown that all stress components vary with the distance *r* to the dislocation as:

$$\blacktriangleright \qquad \sigma \propto \frac{\mu b}{r} \qquad\qquad (2.55)$$

where μ is the elastic shear modulus and *b* the length of the Burgers vector.

It follows that a crystal containing dislocations has an internal energy

Table 2.2. *Grain-boundary diffusion*

Crystal	$T(^\circ C)$	D_{0gb} (cm^2/s)	ΔH_{0gb} (kcal/mol)	ΔH (kcal/mol)
Ag	50–500	0.03	20.2	45
Fe	950–1100	2	40	65
Pb	200–260	0.81	15.7	26

From Martin & Perraillon (1980).

[†] The internal strain is the source of the diffraction contrast making the dislocations visible in transmission electron microscopy.

(stored elastic energy) greater than a dislocation-free crystal. The internal energy per unit length of dislocation in a cylinder of radius R is

$$E = \frac{\mu b^2}{4\pi K} \ln\left(\frac{R}{b_0}\right) \tag{2.56}$$

where $K \approx 1$ is a constant depending on the character of the dislocation (screw or edge) and b_0 is the radius of the 'core' within which the atomic displacements are too important for the elastic theory to be applied (b_0 is of the order of a few b). In a crystal containing a *density* of dislocations ρ (equal to the length of line per unit volume), the stress fields mutually cancel in the average at a distance $R \propto 1/\rho^{1/2}$ which can be used in (2.56).

A stress field exerts a force on a dislocation, normal to the line, causing a loop to expand (thus producing strain) if the force has a component in the glide plane. In the simple case, where the stress field reduces to a shear stress parallel to the Burgers vector of a straight edge dislocation, the force on the

Fig. 2.4. Dislocation loop. (*a*) Half a rectangular dislocation loop is shown in a crystal schematized by a stack of lattice planes (vertical). The extra half plane above the edge portion at left is seen to wind itself helicoidally around the screw portion (back) until it ends up as the extra half plane below the edge portion at right. Each 'turn of the screw' corresponds to a displacement by the Burgers vector **b**, parallel to the screw portion and normal to the edge portions. The 'slipped' area is stippled.

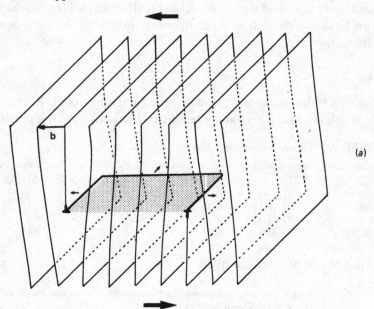

(*a*)

dislocation is parallel to the Burgers vector and its magnitude F is easily computed by writing that the work done by the shear stress σ during slip over a distance b is equal to the work done by the force F on the dislocation over the length of the crystal (producing the same slip); one finds that

▶ $$F = \sigma b \qquad (2.57)$$

A segment of dislocation pinned at its extremities bows out under stress, with a radius of curvature proportional to $\mu b/\sigma$; under a given stress, segments longer than a critical length

▶ $$l_c = \frac{2\mu b}{K\sigma} \qquad (2.58)$$

can go through the semi-circular unstable configuration and emit expanding dislocation loops, while the segment springs back to the initial position to bow out again and so forth (fig. 2.6). Operation of these 'mills' or

(b) Front view of the half-loop. Flipping the bond *OM* to *ON* makes the edge portion propagate towards the left (in the sense of **b**). (c) Top view of the whole loop, oriented continuously (dashed arrows); opposite-sign extra half planes correspond to Burgers vectors of edge portions that would be in opposite sense if the dislocation lines were oriented in the same sense (i.e. not in continuity).

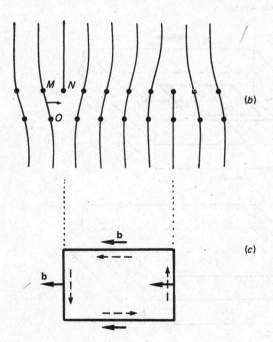

Frank–Read sources is the principal process through which dislocations are created and renewed during deformation of crystals.

Dislocations moving through the crystal interact elastically: a dislocation moving in the stress field of others feels a force and vice versa. If the dislocations attract but lie in parallel glide planes they tend to form *walls* (tilt walls in the case of edge dislocation with the same Burgers vector) (fig. 2.7); if they can cross, they react to form stable *junctions* (fig. 2.8) whose Burgers vector is such that the geometric sum of the Burgers vectors at a triple node is equal to zero. If the dislocations repel but are forced together by the stress, they eventually cross, each leaving on the other a *jog* equal in length and direction to its own Burgers vector (fig. 2.9). Crossing with formation of jogs can be helped by thermal agitation, whereas it is energetically very difficult to destroy a junction.

Reactions between dislocations leading to junctions and eventually to *tangles*, thus greatly impeding glide of other dislocations, are one of the most important processes of strain hardening.

Fig. 2.5. Propagation of slip: the widening of the dislocation loop under stress causes the slipped area to increase until the dislocation moves out of the crystal leaving a step on the surface equal in length to its Burgers vector.

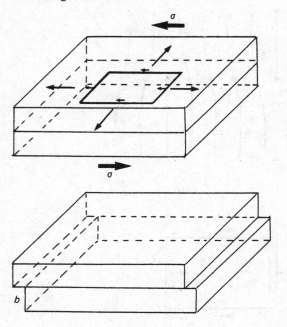

Fig. 2.6. Frank–Read source. (*a*) Dislocation segment of length *l*, stress σ increases with *t*, the source is activated for σ > σ$_c$. (*b*) Constant stress σ, distribution of segments of various lengths, the segments of length *l* > *l*$_c$ are activated.

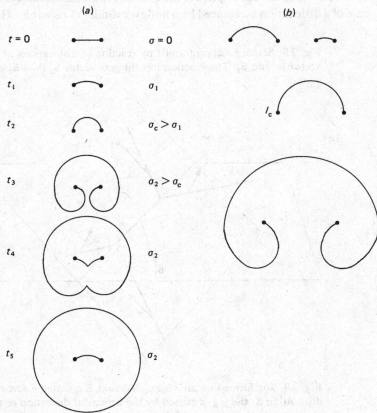

Fig. 2.7. Tilt wall: Stable array of edge dislocations with the same Burgers vector in parallel planes.

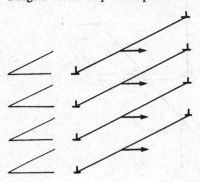

2.3.2   Dislocation cores, Peierls stress and splitting

The dislocation core is the region close to the dislocation line (a few Burgers vectors wide) where the crystalline order is highly defective; the elastic theory cannot be applied in the core and in all elastic calculations the core of a dislocation is replaced by a hollow cylinder of radius b_0. However,

Fig. 2.8. Stable junction formed by reaction of dislocations of Burgers vector $\mathbf{b}_1$ and $\mathbf{b}_2$. The junction has Burgers vector $\mathbf{b}_3$ ($\mathbf{b}_1 + \mathbf{b}_2 + \mathbf{b}_3 = 0$).

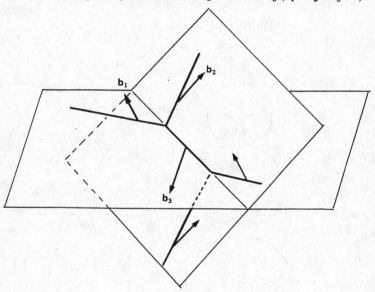

Fig. 2.9. Jog formed on an edge dislocation E crossing a screw dislocation S: the jog is caused by the helicoidal distortion of the glide plane due to the presence of the screw.

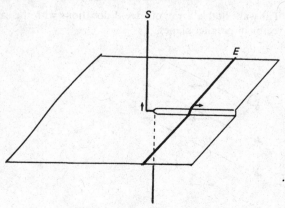

this approximation, perfectly valid as long as we want to determine the long-range stress field, breaks down if we are interested in the mobility of the dislocation. Indeed, it is within the core that the bonds are broken and that the processes controlling the propagation of the dislocation take place. If we consider a dislocation line lying along a crystallographic direction, in a potential trough, it must overcome an energy hill to move into the neighbouring potential trough. The stress necessary to move the dislocation over the hills can be computed in the framework of a dislocation model in a periodic lattice (the Peierls–Nabarro dislocation). The 'Peierls force' is the maximum value of the derivative with respect to the distance of the energy of the Peierls dislocation W and reflects the lattice friction force on the dislocation; the 'Peierls stress' is related to the force by (2.57). It can be shown that

$$\sigma_P = \frac{1}{b}\frac{dW}{dx}\bigg)_{max} = \frac{2\mu}{K}\exp\left(-\frac{4\pi\zeta}{b}\right) \tag{2.59}$$

where ζ is the width of the core, spread on the glide plane, and proportional to the distance d between planes. It is then clear that dislocations glide more easily on close-packed planes (with large d) because their cores are more widely spread in them, which leads to a lower Peierls stress (or lower lattice friction).

The overcoming of a potential hill occurs progressively by lateral spreading of *double kinks* (fig. 2.10) whose formation is thermally activated.

In certain structures the cores of screw dislocations can be spread on several planes simultaneously; the dislocations are then straight and prevented from gliding on one plane by the spreading on others (they are sessile): glide can only occur if the applied stress, helped by thermal agitation, causes the spreading to concentrate on one plane (this is the case for BCC metals at low temperatures).

In many structures whose Bravais lattice is not simple, the spreading of the core can take an extreme form and may be analysed as a ribbon of *stacking fault* bounded by two *partial dislocations* whose Burgers vectors are not Bravais lattice vectors but geometrically add up to the Burgers vector of the total dislocation. This is called *splitting* or dissociation of dislocations (fig. 2.11). (See fig. 2.12 for the case of FCC crystals.) The stacking fault is a planar defect corresponding to a 'fault' in the normal stacking of the structure (e.g. in FCC crystals the stacking ABCABC ... may become ABCACABC ...) and possesses a specific energy γ_F.

The partial dislocations repel elastically and tend to increase the width of the ribbon of stacking fault at a cost of energy, until an equilibrium width is reached, which is inversely proportional to the stacking-fault energy. Split screw dislocations are thus confined to the plane where they are split and can change glide planes (or *cross-slip*) only if the stacking-fault ribbon recombines over a certain length to allow the bowing out of the dislocation on the other plane (fig. 2.13). This process, thermally activated, results in wavy slip lines at the scale of the optical microscope.

2.3.3   Climb of edge dislocations

An edge dislocation can move in the plane normal to its glide plane by a slow process (*climb*) involving transport of matter by diffusion to or from the extra half plane. Climb occurs progressively by migration of jogs on the dislocation line through emission or absorption of vacancies (fig. 2.14).

A dislocation line climbs with a velocity

$$v_c = C_j v_j \tag{2.60}$$

where C_j is the concentration of jogs of height and length b (number of jogs per unit length) and v_j the migration velocity of jogs. Consider, for instance,

Fig. 2.10. (*a*) Peierls potential troughs corresponding to dense crystallographic direction. A dislocation aligned in the bottom of a trough is pushed forward by the applied stress (with a force $F = \sigma b$) and dragged backwards by Peierls stress with a force $f = -dW/dx$. The dislocation can overcome the hill between two troughs by spreading apart of two kinks (*b*).

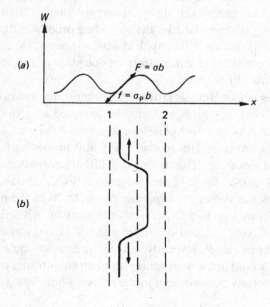

the case of an undissociated dislocation where the jogs migrate by absorption of vacancies. If ϕ vacancies arrive per second on the unit length of dislocation line and are immediately absorbed, each jog of length b receives ϕb vacancies and migrates by b for each vacancy, i.e.

$$v_{\mathrm{j}} = \phi b^2 \tag{2.61}$$

Assuming $C_{\mathrm{j}} \approx 1$ (dislocation 'saturated' with jogs), we have:

$$v_{\mathrm{v}} = \phi b^2 \tag{2.62}$$

Now, ϕ is related to the flux of vacancies J through a cylinder of radius b,

Fig. 2.11. Splitting of dislocations in talc into four partials separated by stacking-fault ribbons. Transmission electron microscopy. Scale bar = 1 μm (courtesy P. Delavignette).

surrounding the dislocation, by

$$\phi = 2\pi bJ \qquad (2.63)$$

and J is given by Fick's law (2.19):

$$J = -D_v \operatorname{grad} C_v \qquad (2.64)$$

The concentration gradient of vacancies between the dislocation and the bulk of the crystal is due to the fact that the applied stress σ modifies the

Fig. 2.12. Splitting of edge dislocations in the FCC structure. (*a*) Top view of three dense (111) planes in position ABC. The Burgers vector is $\frac{1}{2}[110]$ and the corresponding (110) extra half plane is double (bracket). Slip of one Burgers vector involves moving in two steps $\mathbf{b}_1$ and $\mathbf{b}_2$. (*b*) Split dislocation: the first partial has Burgers vector $\mathbf{b}_1$, the second has $\mathbf{b}_2$. They are separated by a stacking-fault ribbon ABCACABC. (*c*) Side view of an undissociated $\frac{1}{2}[110]$ dislocation. Note the double extra $\{110\}$ half planes. (*d*) Splitting of the dislocation ($\frac{1}{2}[110] \rightarrow \frac{1}{6}\{121\} + \frac{1}{6}[21\bar{1}]$) is achieved by pushing apart the two halves of the double extra half plane producing the stacking fault ABCACABC.

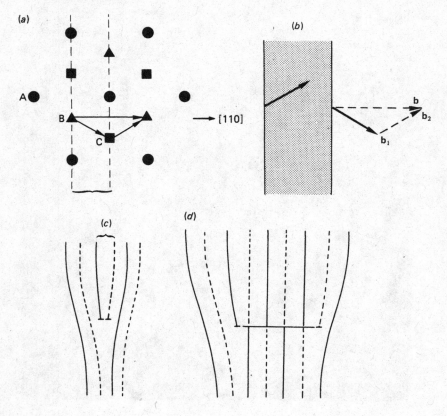

equilibrium concentration of vacancies at the dislocation by increasing the formation energy of vacancies by $\sigma\Omega$, the work done against the stress during the formation of the vacancy ($\Omega \approx b^3$ is the atomic volume).

$$C_v = \frac{1}{\Omega} \exp\left(-\frac{\Delta G_f + \sigma\Omega}{kT}\right) = C_0 \exp\left(-\frac{\sigma\Omega}{kT}\right) \tag{2.65}$$

where C_0 is the concentration of vacancies in the bulk, far from the dislocation, at a distance λ.

Hence

$$\text{grad } C_v \approx \frac{C_v - C_0}{\lambda} \approx \frac{C_0}{\lambda}\left[\exp\left(-\frac{\sigma\Omega}{kT}\right) - 1\right] \tag{2.66}$$

and if $\sigma\Omega \ll kT$

$$\text{grad } C_v \approx -C_0 \frac{\sigma\Omega}{kT\lambda} \tag{2.67}$$

Fig. 2.13. Cross-slip: a screw dislocation split in plane P_1 slips in it. To cross-slip onto plane P_2, the stacking-fault ribbon must locally constrict. The constricted portion develops in P_2 where it can split again.

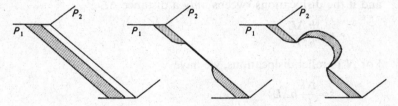

Fig. 2.14. Climb of an edge dislocation: the dislocation line climbs by an interatomic distance when a jog travels along its length by absorbing (A) or emitting (E) vacancies.

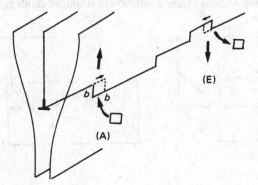

with equations (2.60) to (2.67) we obtain

$$v_c \approx 2\pi b^3 D_v C_0 \frac{\sigma\Omega}{kT\lambda} \tag{2.68}$$

or with $b^3 D_v C_0 = D_v N_v = D_{sd}$ (2.13)

▶ $$v_c \approx \frac{2\pi}{\lambda} D_{sd} \frac{\sigma\Omega}{kT} \tag{2.69}$$

2.3.4   Orowan's equation

Orowan's equation relates the strain-rate caused by glide (or climb) of dislocations to the density, strength and velocity of the mobile dislocations. It is in fact a microscopic constitutive equation which lies at the basis of most equations describing the various deformation processes.

Consider a crystal in the shape of a parallelepiped with dimensions h, l, L (fig. 2.15) and one straight edge dislocation, of length l and Burgers vector b, sweeping its glide plane over the distance L (or a screw dislocation of length L, sweeping the same glide plane over l); when the dislocation has swept across the whole crystal, the average shear strain is

$$\varepsilon = \frac{b}{h}$$

and if the dislocations sweeps only a distance ΔL

$$\varepsilon = \frac{b}{h} \frac{\Delta L}{L}$$

For N parallel dislocations, we have

$$\varepsilon = \frac{Nl}{V} b\Delta L$$

Fig. 2.15. Orowan's equation. (*a*) A straight edge dislocation, sweeping its glide plane over ΔL creates shear strain $\varepsilon = b\Delta L/hL$. (*b*) Two straight edge dislocations create a pure shear strain by climbing in opposite senses.

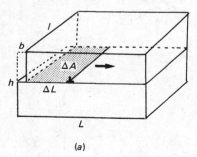
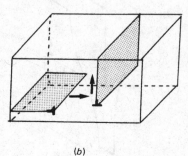

(*a*)   (*b*)

where $V = Llh$ is the volume of the crystal, or

$$\varepsilon = \rho b \Delta L \tag{2.70}$$

with $\rho = Nl/V$, density of mobile dislocations.

The same expression holds for more realistic situations (curved dislocations on various glide planes) taking ρ and ΔL as average values.

The strain-rate is then:

$$\dot{\varepsilon} \equiv \frac{d\varepsilon}{dt} = b\frac{d}{dt}(\rho \Delta L) = b\left[\Delta L\frac{d\rho}{dt} + \rho\frac{d\Delta L}{dt}\right] \tag{2.71}$$

In most cases, it is reasonable to assume that the dislocation density does not vary much at the time scale considered and eq. (2.71) can be written

▶ $$\dot{\varepsilon} = \rho b v \tag{2.72}$$

where v is the average velocity of the dislocations. This expression can be used for quasi-steady-state creep, but obviously not in cases of sigmoidal creep or when there is a yield point (see §1.1.3), since the rapid increase in strain-rate (or decrease in stress) is due in those cases to a sudden increase in the dislocation density.

Orowan's equation (2.72) is clearly a transport equation as mentioned above (§2.1.1); it can be cast in the shape of a microscopic constitutive relation by expressing the velocity and the density as functions of the applied stress, temperature and pressure.

Shear strain can also be produced by climb of an equal number of edge dislocations of opposite sign exchanging vacancies at constant total volume: the pure shear strain resulting from the increase in area of the extra half planes of one family at the expense of those of the other (fig. 2.15(*b*)) can be put in the form of (2.70); Orowan's equation is valid whether the dislocations move by glide or by climb.

2.4 Grain boundaries and recrystallization

2.4.1 Definitions

A grain boundary is a two-dimensional lattice defect that introduces a misorientation in the lattice with no long-range stress field. As a consequence, a single crystal containing a *high-angle grain boundary* is best described as two crystals differing in orientation, on each side of the grain boundary; a low-angle grain boundary is, however, seen as a defect in a single crystal.

A grain boundary is characterized by three parameters describing the relative orientation of the lattices on each side of it as the result of a rotation about an axis (two parameters to define the rotation axis + 1 for the angle) and two parameters fixing the orientation of the boundary itself in one of the lattices (altogether five degrees of freedom).

A *twist-boundary* is normal to its axis of rotation and a *tilt boundary* contains its axis of rotation. A general boundary has both tilt and twist components (as a general dislocation has edge and screw components).

The general analysis of grain boundaries has been done in considerable detail in the case of cubic lattices (see Bollmann, 1970; Priester, 1980); although many of the concepts introduced do not apply directly to the case of non-cubic lattices, the general results derived remain valid on the whole. In cubic lattices, it is useful to introduce the *Coincidence Site Lattice* (CSL) or lattice of the sites common to both lattices. The degree of coincidence of the lattices of each grain is expressed by the twin index Σ equal to the inverse of the proportion of sites in coincidence (e.g. for cubic crystals $\Sigma = 5$ for a rotation of $37°$ about $[100]$ (fig. 2.16)). A *coincidence boundary* contains coincidence sites. It is also useful to introduce the *Displacment Shift Complete Lattice* (DSCL), whose lattice vectors are the sums and differences of lattice vectors of the two adjoining crystals.

2.4.2   Grain-boundary dislocations

The lattice misorientation introduced by grain boundaries can formally be expressed as the misorientation introduced by a planar array of dislocations. Indeed, low-angle grain boundaries (or subgrain boundaries) are dislocation walls whose dislocations can be seen by transmission electron microscopy (see chapter 6). The misorientation θ is inversely

Fig. 2.16. Coincidence site lattice for a cubic crystal: rotation $\theta = 37°$ about $[100]$, $\Sigma = 5$. Lattice A – circles; lattice B – triangles; coincidence site lattice–squares. The DSCL is drawn (thin lines) in the lower left quadrant.

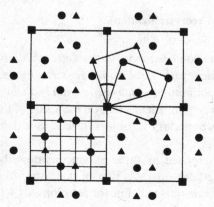

proportional to the dislocation spacing in the sub-boundary. For a tilt wall (see fig. 2.17):

▶ $$\theta = \frac{b}{d} \qquad (2.73)$$

where d is the spacing of edge dislocations and b their Burgers vector. Twist walls are arrays of screw dislocations and are related to tilt walls much in the way screw dislocations are related to edge dislocations. Indeed, a small parallelepiped inside a crystal, misoriented by rotation about an axis parallel to one of its edges is bounded by four tilt walls and, necessarily, two twist walls (fig. 2.18).

There is obviously no clear-cut distinction between low-angle and high-angle grain boundaries; the conventional limit is usually set at a misorientation of about 10 or 15° which correspond roughly to the dislocation spacing below which the cores would be contiguous. When a grain boundary can be analysed in terms of dislocations, these are called *intrinsic grain-boundary dislocations* (GBD). The primary intrinsic GBDs are responsible for the misorientation and secondary intrinsic GBDs accommodate small deviations from the nearest coincidence orientation of the lattices or the boundary (Hirth, 1972; Hirth & Balluffi, 1973). The stress fields of the intrinsic GBDs mutually cancel at a short distance from the boundary,

Fig. 2.17. Low-angle tilt boundary made up of parallel edge dislocations of Burgers vector $\mathbf{b}$, distant by d. Misorientation $\theta \approx 2 \tan(\theta/2) = b/d$.

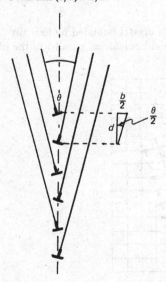

comparable to the dislocation spacing. *Extrinsic* GBDs can be trapped in the boundary, coming from the neighbouring grains.

2.4.3   Energy of grain boundaries

The energy per unit area of a low-angle boundary is easily computed as equal to the energy per unit length of a dislocation (given by (2.56) with $R = d$) multiplied by the number $1/d$ of dislocations per unit length; with (2.73) we obtain:

$$E = \frac{\mu b^2}{4\pi K}\, \theta \left[\ln \left(\frac{b}{b_0} \right) - \ln \theta \right] \tag{2.74}$$

The energy increases with increasing misorientation up to about 15°, when (2.74) ceases to be valid (fig. 2.19). The energy of high-angle boundaries is roughly independent of θ, except for coincidence orientations, where it is lower. For a given lattice misorientation, the grain-boundary energy depends on the orientation of the boundary plane.

2.4.4   Structure of grain boundaries

The structure of grain boundaries on the atomic scale has been the object of active investigation in recent years. Analysis of grain boundaries by X-ray and electron diffraction, as well as the high-resolution transmission electron microscopy (Gronsky, 1979; Hall *et al.*, 1982; Bristowe & Sass, 1980; Krivanek *et al.*, 1977), has finally dealt the last blow to the old idea that grain boundaries were an amorphous layer between grains: crystalline order persists on each side of the boundary up to a defective core

Fig. 2.18. Rotated block inside a crystal bounded by four tilt boundaries (*ABCD*) and two twist boundaries, parallel to the plane of the page.

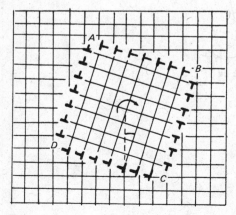

region, one or two atomic distances thick, which can be analysed using several convergent approaches. In what we can call a crystallographic approach, the grain boundary is described by a 'fit–misfit' model (Balluffi *et al.*, 1982); in the regions of 'fit' there is partial lattice matching across the boundary and the regions of 'misfit' are line defects with dislocation and/or boundary step character, described by using the concept of DSC lattice. The structure of the core can also be described as a packing of deltahedra (fig. 2.20) (polyhedra with equilateral triangles as faces); this approach is useful to understand the segregation of impurities to the boundary; it is compatible with the description of low-angle boundaries as dislocation walls since dislocation cores can also be seen as composed of deltahedra (Ashby *et al.*, 1978). Computer modelling has been widely used in metals for which reasonably good interatomic pair potentials are known (Vitek *et al.*, 1979; Bristowe & Sass, 1980). Molecular dynamics have been used for FCC solid rare gases; a structural transition towards a partially disordered structure is found at high temperature (Cicotti *et al.*, 1983).

In ceramic oxides and minerals, the problem of the boundary structure is much more complicated than in metals owing to the existence of several electrostatically interacting charged species (Balluffi *et al.*, 1981).

Fig. 2.19. Dependence of the relative grain-boundary energy on misorientation θ: nickel oxide, 1520 °C, symmetrical tilt boundaries (after Dhalenne *et al.*, 1982).

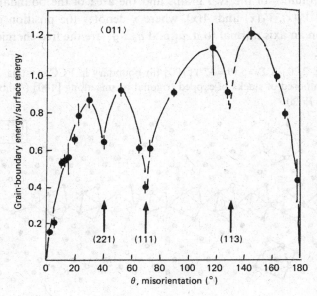

2.4.5   Grain-boundary migration and sliding

Grain boundaries can move in their plane (*sliding*) or normal to it (*migration*).

Grain-boundary sliding (sometimes over large distances) occurs under applied shear stress. It is an important strain-producing process in superplastic deformation (chapter 8) and it is usually evidenced by the offset of a fiducial grid scribed on the specimen. Only high-angle boundaries usually slide, as the dislocations in low-angle boundaries are generally free to respond to the stress by moving in their own glide planes. Sliding can be explained in terms of climb and glide of grain-boundary dislocations, which may be the cause of some accompanying migration of the boundary (Pond et al., 1978). Migration of grain boundaries can also occur under stress and produce strain: it has been documented in creep experiments on aluminium (Exell & Warrington, 1972) and NaCl (Guillopé & Poirier, 1979) and can be accounted for by the propagation of ledges (steps) associated to extrinsic grain-boundary dislocations (Guillopé & Poirier, 1980). When a tilt grain boundary migrates under stress, it leaves behind a slope or talus at the surface of the crystal, corresponding to the shear caused by the motion of the component dislocations (fig. 2.21).

There are, however, other driving forces than applied stress for the migration of grain boundaries and they are mostly responsible for the important phenomenon of recrystallization (see §2.4.7).

Let us consider a grain boundary separating two crystals 1 and 2 (fig. 2.22). The volumes of the two grains and the area of the boundary are respectively $V_1(x)$, $V_2(x)$ and $A(x)$, where x denotes the position of the boundary on an axis normal to it, g_1 and $g_2 > g_1$ are the free energies per

Fig. 2.20. A $\Sigma = 5$ ($\theta = 37°$) [100] tilt boundary in FCC crystals composed of stacks of capped trigonal prisms along [100] (Ashby *et al.*, 1978).

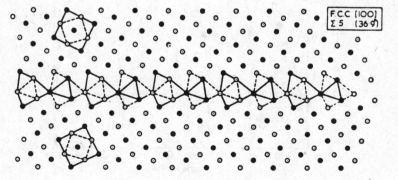

unit volume of grains 1 and 2. Motion of the boundary by dx in the direction of 2 will replace a volume dV of grain 2 by the same volume of grain 1, while the area of the boundary changes by dA. There is a net change in total free energy dG and the driving force is

$$F = -\frac{dG}{dx} = (g_2 - g_1)\frac{dV}{dx} - g_{gb}\frac{dA}{dx} \tag{2.75}$$

where g_{gb} is the free energy per unit area of the grain boundary. We see that

Fig. 2.21. Grain-boundary migration and sliding: NaCl crystals cleaved along (100) plane (scale bar=0.1 mm). (*a*) Grain-boundary migration under stress. The grain boundary migrating towards the right leaves a sloping talus behind. (*b*) Grain-boundary sliding leaves a step on the surface. The boundary has migrated towards the left inside the crystal after sliding took place (courtesy M. Guillopé).

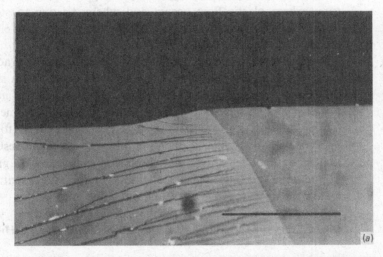

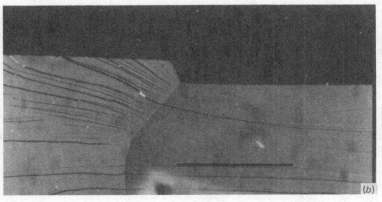

the driving force for migration under no applied stress can be decomposed into a volume driving force and a surface driving force. The difference in specific free energy $\Delta g = g_2 - g_1$ may have various physical sources:

  (i) A difference in dislocation density between the grains. The grain with the higher density, hence the higher free energy, is consumed by the dislocation-free grain, dislocations are absorbed by the boundary (Pond & Smith, 1977). This is the basis of *strain-induced boundary migration* (SIBM) in annealed deformed crystals, a fundamental process in primary recrystallization.

 (ii) A difference in elastic strain energy (due to the anisotropy of elastic constants) between the two grains differently oriented with respect to an applied stress that causes only elastic deformation (Kamb, 1961). This is the basis of the so-called stress-induced recrystallization, which, in most cases, is probably negligible at high temperatures where plastic flow occurs and where Δg due to stored dislocations is much greater than Δg due to anisotropy of elastic energy. (See Paterson, 1973.)

(iii) A difference in chemical potential of one of several species across the grain boundary in the case of alloys or minerals with slight differences in chemical composition. *Diffusion-induced grain-boundary migration* (DIGM) has been recently documented (Hillert & Purdy, 1978) in thin pure iron specimens exposed to zinc vapour at 600 °C; transport of zinc into the specimen resulted almost entirely from boundary diffusion coupled with grain-boundary sweeping; all the original volume was eventually transformed into a recrystallized high-zinc alloy. This phenomenon has been actively investigated in recent years (Shewmon, 1981; Balluffi & Cahn, 1981; Chongmo & Hillert,

Fig. 2.22. Grain-boundary migration. The grain boundary migrates by dx from position 1 to 2; it increases its area by dA and the volume of the left-hand-side grain increases by dV at the expense of the right-hand-side grain.

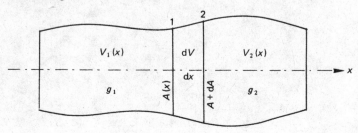

1982). The possibility of a chemical driving force for boundary migration in mica has been suggested by Etheridge & Hobbs (1974).

In most of the cases of boundary migration, it is possible to find a mechanism for migration involving motion of grain-boundary dislocations and of their associated steps (Rae & Smith, 1980; Balluffi & Cahn, 1981).

The surface driving force leads to minimization of the total grain-boundary area. It is responsible for secondary recrystallization, or more properly grain growth, a stage which follows primary recrystallization and during which the dislocation-free new grains reach an equilibrium structure, the smaller grains being consumed by the bigger ones.

2.4.6 Mobility of grain boundaries

The velocity of migrating grain boundaries has been investigated under various conditions and driving forces (see Smith *et al.*, 1979). The relation between velocity v and driving force F is usually written:

$$v = MF \tag{2.76}$$

where M is the mobility.

The mobility increases with temperature, and may vary with the misorientation, the degree of coincidence, etc., but there is so far no consistent pattern and it is not possible to make general statements.

The effect of small concentrations of impurities on the mobility of grain boundaries may be considerable; it has been investigated by Lücke & Stüwe (1971). Impurity atoms are attracted to the defective core of a grain boundary and at a boundary at rest the concentration of impurity atoms C is higher than the bulk concentration C_0:

$$C(x) = C_0 \exp\left(-\frac{U(x)}{kT}\right) \tag{2.77}$$

where x is the distance to the boundary and $U(x)$ the interaction energy. For steady-state motion of the grain boundary, the impurity atoms diffuse into a moving potential well and the diffusion flux of atoms following the boundary is equal to the advective flux of atoms toward the moving boundary; a steady-state concentration of impurities at the boundary, $C(x, v, C_0)$, sets up, which depends on the velocity v of the boundary. The total drag force exerted by impurity atoms on the boundary is:

$$f(v, C_0) = \int_{-\infty}^{+\infty} C(x, v, C_0)\frac{dU(x)}{dx}dx \tag{2.78}$$

The grain-boundary velocity is therefore:

$$v = M[F - f(v, C_0)] \tag{2.79}$$

with

$$M \propto \exp\left(-\frac{Q}{kT}\right)$$

The drag force is equal to zero for zero and infinite velocity and it can be qualitatively represented by a bell-shaped curve. Eq. (2.79) can then be graphically solved (fig. 2.23). For concentration higher than a critical value C_c, two velocity regimes are possible at constant temperature (i.e. M = const):

- at low driving forces F, the velocity is low, the grain boundary is 'loaded' with impurities and its velocity is controlled by the diffusion of impurities;
- at high driving forces, the grain boundary can tear itself from the

Fig. 2.23. Grain-boundary mobility. The grain-boundary velocity v is found by graphically solving the equation that expresses the velocity v as a function of the driving force F, the mobility M and the impurity drag force that in turn depends on v; the solutions are found at the intersection of the bell-shaped curve representing the drag force $y = f(v)$ with the straight line $y = F - v/M$. There is one solution for $F < F_2$ and $F > F_1$ and two unstable solutions for $F_2 < F < F_1$, for concentration above a critical value C_c. For $C < C_c$ there is only one solution.

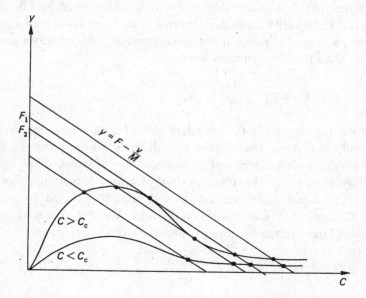

impurities and the free boundary moves faster, with its intrinsic velocity.

The surface $v(F, C)$ can be seen as a 'cusp catastrophe' surface (fig. 2.24) (Woodcock & Davis, 1978).

2.4.7 Static recrystallization

Static recrystallization is a modification of the grain structure (size, shape, orientation of grains) that occurs during high-temperature annealing following deformation (dynamic recrystallization occurs during deformation; it will be treated in chapter 6).

Recrystallization involves grain-boundary migration and the driving forces for recrystallization are the same as for grain-boundary migration (§ 2.4.5).

Primary recrystallization is driven by the free energy difference between deformed grains with a high dislocation density and dislocation-free grains; it can be described with the same formalism as phase transformations by nucleation and growth. It follows from eq. (2.75) that above a critical size a

Fig. 2.24. Cusp catastrophe set for the velocity of the grain boundary as a function of driving force F and impurity content C. For high C and low F, the boundary is loaded with impurities (low v); for low C and high F, the boundary is free (high v).

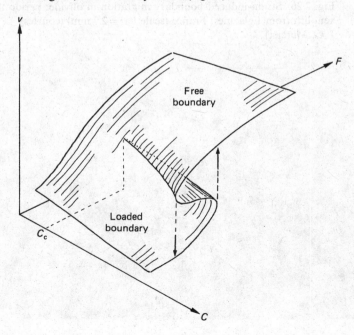

Fig. 2.25. Strain-induced boundary migration in aluminium ($\varepsilon = 40\%$, 1 h anneal at 328 °C). Each grain locally bulges into its neighbour. (Scale bar = 50 μm) (courtesy R. Doherty.)

Fig. 2.26. Strain-induced boundary migration in olivine: peridotite xenolith from Eglazines, France (scale bar = 2.7 mm) (courtesy J. C. Mercier).

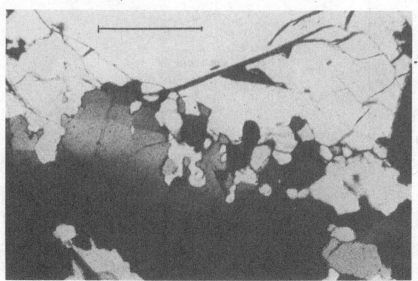

dislocation-free nucleus can spontaneously grow, resulting in a decrease of the total free energy. Experimental evidence supports the idea that nuclei are not formed by statistical fluctuations but can be identified as local bulges on high-angle grain boundaries: any kind of boundary formed during deformation can locally bow out if its misorientation is high enough and the bulge can grow by strain-induced boundary migration (SIBM) (fig. 2.25) (Bailey & Hirsch, 1962). The high-angle grain boundaries may be kink-band or twin boundaries, e.g. in galena (McClay & Atkinson, 1977), enstatite (Etheridge, 1975), mica (Etheridge & Hobbs, 1974) or olivine (fig. 2.26). SIBM also occurs at the boundaries of deformation bands, highly misoriented regions in a crystal due to local divergence in the sense of rotation of slip planes during deformation (Bellier & Doherty, 1977; Inokuti & Doherty, 1978).

When all deformed grains have been swept by grain boundaries and cleaned of most dislocations, the polycrystal may still lower its free energy by reducing the total grain-boundary area (Deeley, 1895). This is achieved by *grain growth*, which results in an equilibrium polygonal texture or even in single crystals in some cases (Lacroix, 1903).

Readings

R. W. Cahn & P. Haasen, Eds., *Physical metallurgy*, 3rd edn, North-Holland, Amsterdam (1983). All physical metallurgy in 2000 pages.

J. Weertman & J. R. Weertman, *Elementary dislocation theory*, Macmillan, New York (1964). A clear and concise textbook.

J. Friedel, *Dislocations*, Pergamon (1964), 491 pp. Rich and full of physical insight.

American Society for Metals, *Grain boundaries, structure and kinetics*, ASM, Metals Park, Ohio (1979), 470 pp. The proceedings of an international conference, in the shape of a collective treatise by the foremost authorities in the field. A state of the art review on grain boundaries.

3

Phenomenological and thermodynamical analysis of quasi-steady-state creep

The phenomenological parameters expressing the dependence of creep rate on temperature and stress are introduced. The apparent activation energy may vary with temperature (curved Arrhenius plot) when several 'parallel–concurrent' or 'series–sequential' processes operate. The creep-rate–stress logarithmic plot is usually curved over a large stress range: at low stresses it can be assimilated to portions of straight lines (power-law creep) but this fails at high stresses when the stress dependence of creep-rate may be exponential, reflecting a stress dependence of the apparent activation energy.

Thermally activated processes of dislocation motion are analysed in the light of the theory of absolute reaction rates. At an activation site (flow unit) a dislocation segment has to overcome an energy barrier to contribute an elementary strain. The energy to overcome the barrier (visualized by the force–distance diagram) is provided by the applied stress and the thermal agitation. The thermodynamic activation parameters are defined and their experimental determination is discussed.

3.1   Phenomenology

3.1.1   Generalities

Phenomenological analysis of high-temperature deformation is best done on quasi-steady-state, constant-stress creep-test results, since they directly provide time-independent empirical equations of state of the form: $\dot{\varepsilon} = f(\sigma, T, P)$. Phenomenological parameters expressing the dependence of the creep-rate on temperature and applied stress are analysed in the framework of a thermodynamical theory of deformation (see below) and, when possible, compared with the values expected if certain physical processes control the creep-rate. (Note that the process that controls the creep-rate $\dot{\varepsilon}$, i.e. the dislocation average velocity in Orowan's equation (2.72), is not necessarily the same as the one that produces the strain.)

The creep-rate at constant stress usually increases exponentially with temperature (Andrade, 1910, 1914; Chevenard, 1922); it is therefore convenient to plot $\ln \dot{\varepsilon}$ against $1/T$, thus fitting the creep-rate to an Arrhenius-type law:

$$\dot{\varepsilon} \propto \exp\left(-\frac{Q}{RT}\right)$$

where Q, proportional to the slope of the $\ln \dot{\varepsilon}$–$1/T$ plot is the *apparent activation energy* of creep.

The creep-rate increases with the applied stress σ and it is usual to plot $\log \dot{\varepsilon}$ against $\log \sigma$ in the hope of obtaining a straight line by virtue of the log–log plot. When this happens, and it happens most of the time if the stress interval is small enough, the creep-rate depends on the stress by a *power law* $\dot{\varepsilon} \propto \sigma^n$: n is the *stress exponent*; it is the inverse of the strain-rate sensitivity of the stress defined by eq. (1.37):

$$n = \frac{1}{m} = \frac{d \ln \dot{\varepsilon}}{d \ln \sigma}$$

The empirical mechanical equation of state can then be written:

▶ $$\dot{\varepsilon} = \dot{\varepsilon}_0 \sigma^n \exp\left(-\frac{Q}{RT}\right) \tag{3.1}$$

This equation has been proposed for creep by Dorn (1956). Zener & Hollomon (1944) suggested the use of the parameter

$$Z = \dot{\varepsilon} \exp\left(\frac{Q}{RT}\right) \tag{3.2}$$

to investigate the strain-rate dependence of the stress in constant strain-rate tensile experiments; the parameter Z or Zener–Hollomon parameter has, in fact, been widely used to investigate the stress dependence of the creep-rate, using experiments at various temperatures.

It may be helpful to visualize the temperature and stress dependence of the strain-rate by a three-dimensional plot using as coordinate axes: $x = 1/T$, $y = \ln \sigma$, $z = \ln \dot{\varepsilon}$. The mechanical equation of state is then represented by a surface $\ln \dot{\varepsilon} = f(1/T, \ln \sigma)$ (fig. 3.1), whose intersections by planes $y = \text{const.}$ and $x = \text{const.}$ represent the variation of strain-rate with temperature (Arrhenius plot) and applied stress respectively. The domain of experimental conditions (σ and T) can be delineated in the plane $z = \text{const.}$ Projection of the sections onto the coordinate planes followed by rotation of the planes about the x and y axes provides a convenient way of physically correlating the Arrhenius plot and stress–strain curves (fig. 3.1(b)) (Poirier, 1976).

We have represented in fig. 3.1 the most general case, where the

Fig. 3.1. (*a*) Representation of the mechanical equation of state by a surface ln $\dot{\varepsilon} = f(1/T, \ln \sigma)$. (*b*) Projection of the surface on the coordinate planes: (ln $\dot{\varepsilon}$, $1/T$) – Arrhenius plots at constant stress; (ln $\dot{\varepsilon}$, ln σ) – σ, $\dot{\varepsilon}$ plots at constant temperature; (ln σ, $1/T$) – experimental conditions.

(*a*)

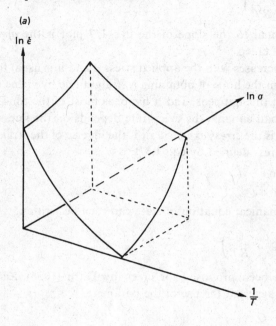

(*b*)

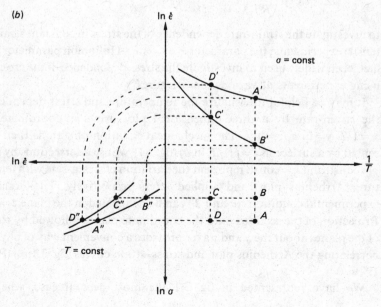

$(\ln \dot{\varepsilon}, 1/T, \ln \sigma)$ surface is not a plane, even though it usually can be assimilated to a plane in a small interval of temperature and stress (i.e. the Arrhenius plot is a straight line with slope $-Q/R$ and the logarithmic strain-rate–stress curve is a straight line with slope n). Experiments over wider intervals frequently yield curved curved Arrhenius plots (temperature-dependent apparent activation energy) and curved strain-rate–stress plots (stress-dependent n). Furthermore, it may be that for a given temperature the apparent activation energy decreases as stress increases, which corresponds to n decreasing as temperature increases at constant stress (fig. 3.1(b)); we will see the significance of the stress-dependent activation energy in §3.2.

3.1.2 Curvature of the Arrhenius plot

If the strain-rate is controlled by a unique, thermally activated process, the Arrhenius plot is a straight line over the whole temperature range investigated and the apparent activation energy of creep is equal to the activation energy of the controlling process. However, it is often the case that several potential rate-controlling processes exist, with different activation energies, with the result that the Arrhenius plot is curved in the temperature range where the activity of the mechanisms is comparable. Two cases must be distinguished (Gifkins, 1970; Langdon & Mohamed, 1977):

(i) The processes are *independent* or *parallel–concurrent*, they are simultaneously active and each contributes (directly or indirectly) a strain ε_i, the strain-rates are additive, thus *in a rheological diagram, parallel–concurrent processes must be represented by dashpots in series*. The total strain-rate is given by:

$$\dot{\varepsilon} = \sum_i \dot{\varepsilon}_i$$

For the simpler (and rather frequent) case of two parallel–concurrent processes the temperature dependence of the creep-rate is given by:

$$\dot{\varepsilon}(T) = \dot{\varepsilon}_{01} \exp\left(-\frac{Q_1}{RT}\right) + \dot{\varepsilon}_{02} \exp\left(-\frac{Q_2}{RT}\right)$$

The Arrhenius plot is the sum of two exponentials (fig. 3.2(a)); if $Q_1 > Q_2$, process 1 gives the dominant contribution to the creep rate at high temperatures and process 2 is dominant at low temperatures: in the temperature range where the activity of both processes is comparable, the Arrhenius plot is curved. In any given temperature range, the *faster* process dominates creep.

(ii) The processes are independent and act in *series* (*series–sequential* processes). Process 2 cannot operate until process 1 has taken place and vice versa (mutually accommodating processes); the time lengths during which each process operates are additive, thus the inverses of strain-rates are additive (if each process contributes a strain):

$$\frac{1}{\dot{\varepsilon}} = \sum_i \frac{1}{\dot{\varepsilon}_i}$$

Fig. 3.2. Arrhenius plots for two rate-controlling processes. (*a*) Parallel–concurrent processes, the faster process controls the overall rate. (*b*) Series–sequential processes, the slower process controls the overall rate.

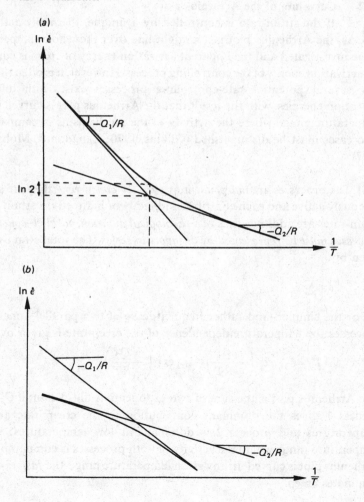

For two series–sequential processes, the temperature dependence of the creep-rate is given by:

$$\dot{\varepsilon}^{-1}(T) = \dot{\varepsilon}_{01}^{-1} \exp\left(\frac{Q_1}{RT}\right) + \dot{\varepsilon}_{02}^{-1} \exp\left(\frac{Q_2}{RT}\right)$$

Here, in any given temperature range, the *slower* process dominates the creep-rate, but not necessarily the creep strain, since the slower process must operate for the faster one to be able to proceed, but it may be that the slower process contributes directly little strain and only allows the other process, with a greater strain contribution, to operate. The process with the higher activation energy is dominant at lower temperatures (fig. 3.2(*b*)).

3.1.3 Curvature of the creep-rate–stress plot

The log $\dot{\varepsilon}$–log σ plot is seldom linear over a wide range of applied stresses, and the power-law equation (3.1) usually does not hold at high stresses. It is, of course, possible to find empirical equations that fit the experimental points over the whole range of applied stresses, but are devoid of any physical meaning. The most successful one has been proposed by Garofalo (1963) and can be conveniently written as the variation with stress of the Zener–Hollomon parameter (3.2):

$$Z = A[\sinh(\alpha\sigma)]^n \tag{3.3}$$

With an adequate choice of constants, such an expression may, in some cases, hold over about 15 orders of magnitude of strain-rates (fig. 3.3) (Jonas, 1969).

Another approach consists in fitting several straight lines, each corresponding to one power law and presumably corresponding to physically meaningful parallel–concurrent processes: at low stresses, processes related to creep by transport of matter by diffusion ($n = 1$ or 2), then recovery-creep processes controlled by diffusion (Weertman creep, see chapter 4) with $n = 3$ to 5; for high enough stresses, thermally activated processes with a stress-dependent activation energy may appear. The creep rate then increases exponentially with stress (§ 3.2) and the log $\dot{\varepsilon}$–log σ plot is no longer a straight line (*power-law breakdown*). Curves in the σ, T experimental conditions planes separate domains where each process is dominant (fig. 3.4).

It must be kept in mind that a power-law exponent n *found by fitting a straight line through experimental points in a small stress range does not necessarily correspond to a power-law creep process* (e.g. Weertman creep) *even if it has a value compatible with it: it may be that a curve corresponding to an exponential dependence on σ can be assimilated to a*

Fig. 3.3. Zener–Hollomon parameter Z as a function of the applied stress: $Z = A\,[\sinh{(\alpha\sigma)}]^n$. Data for aluminium deformed by extension, compression, torsion and tensile creep (Jonas, 1969).

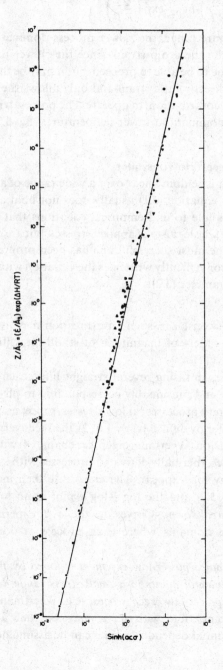

straight line over the stress range considered, thus yielding a spurious power exponent. Indeed, in some cases, it is possible to analyse an experimental log $\dot{\varepsilon}$–log σ curve in terms of one physically meaningful creep law (with a complicated stress dependence) as well as in terms of several power-law regimes. The consequences are not trivial when one wishes to extrapolate the experimental law outside of the experimental stress range, as for instance in the case of olivine, $(Mg, Fe)_2SiO_4$, an important constituent of the Earth's mantle (Poirier & Vergobbi, 1978) (fig. 3.5). Jonas (1969) has clearly shown that the Zener–Hollomon parameter corresponding to the same experimental data can sometimes be convincingly plotted against log σ (power law) or against σ (exponential dependence) with obviously contradictory results for extrapolations.

Fig. 3.4. Composite surface $\ln \dot{\varepsilon} = f(\ln \sigma, 1/T)$ for several creep processes: (1) diffusion creep, $n = 1$; (2) power-law diffusion-controlled creep, $n = 3$, same activation energy as (1); (3) cross-slip-controlled creep.

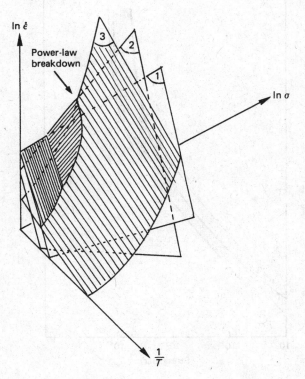

3.1.4   The power-law equation

As we have seen above, the power-law equation describes quite well the creep of a considerable number of materials in the stress range that corresponds to laboratory experimental conditions. We will see in chapter 4 that for many elements (mostly metals) the activation energy for creep is reasonably close to the activation energy for self-diffusion (Sherby & Burke, 1967) (see §4.2 for a discussion of the correlation); the existence of models for diffusion-controlled power-law creep (Weertman, 1968) further justified the introduction of a power-law creep equation under the form proposed by Mukherjee *et al.* (1969) (Dorn equation):

▶ $$\dot{\varepsilon} = A \frac{D\mu b}{kT} \left(\frac{\sigma}{\mu} \right)^n \tag{3.4}$$

Fig. 3.5. Log $\dot{\varepsilon}$–log σ plot for experimental results on creep of olivine normalized to 1400 °C with $Q = 125$ kcal/mol. All data points are within the stippled area and the solid curve is a fit by a cross-slip-controlled law, the dashed curve is a fit by a climb-controlled creep law (after Poirier & Vergobbi, 1978).

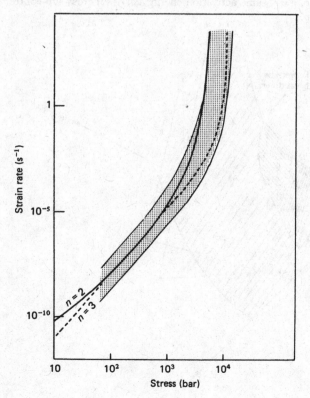

where D is the self-diffusion coefficient, μ the shear modulus and A a dimensionless constant. However, a compilation of many experimental data for various materials (Brown & Ashby, 1980) shows that A and n are not independent:

$$A \approx C_0^{n-2.70} \qquad (3.5)$$

with $C_0 \approx 1025$.

The Dorn equation can now be written in the form known as Norton's law by engineers:

$$\dot{\varepsilon} = Bf(T)\left(\frac{\sigma}{\sigma_0}\right)^n \qquad (3.6)$$

where the stress is normalized to a reference stress $\sigma_0 = \mu/C_0 \approx \mu/1025$, close to the yield stress, rather than to the shear modulus. The coefficient $B = C_0^{-2.70}$ is then truly a constant independent of n.

3.2 Thermodynamics of creep

3.2.1 The absolute reaction-rate theory

The thermodynamical analysis of plastic deformation by dislocation motion stems from the theory of the viscous flow of molecular fluids proposed by Eyring (1936). The flow rate is considered as resulting from the successful operation of 'flow-units', where the strain carrier moves by overcoming an energy barrier with the help of the applied stress and thermal agitation. The formalism is the same as the one used for chemical reactions: the *absolute reaction-rate theory*. Let us consider a segment of dislocation as a flow-unit and assume that the number of times per second that the dislocation segment overcomes the energy barrier in the sense in which the applied stress σ does work can be written as a thermally activated reaction rate:

$$v^+ = v_0 \exp\left[-\frac{Q_0 - \sigma b \Delta A}{kT}\right] \qquad (3.7)$$

where Q_0 is the height of the energy barrier and $\sigma b \Delta A$ is the work done by the applied stress, helping in overcoming the barrier and thus effectively lowering the height of the barrier; ΔA is the area swept in the glide plane while the dislocation segment moves up to the saddle point on top of the barrier (activation area), b is the Burgers vector or the distance between two successive energy troughs, v_0 is the attempt frequency.

The theory rests on the essential assumption that once the barrier has been overcome and the dislocation segment has fallen into the next trough, all its energy is thermalized, it must then acquire anew the total energy to overcome the following barrier as a purely random process.

(i) If the dislocation sits in the trough after overcoming the barrier, it may also jump back, with a frequency:

$$v^- = v_0 \exp\left[-\frac{Q_0 + \sigma b \Delta A}{kT} \right] \tag{3.8}$$

The net forward reaction rate is then

$$v = v^+ - v^- = v_0 \exp\left(-\frac{Q_0}{kT} \right) \sinh\left(\frac{\sigma b \Delta A}{kT} \right) \tag{3.9}$$

which can be linearized if the stress is small, i.e. if $\sigma b \Delta A \ll kT$:

$$v = v_0 \frac{\sigma b \Delta A}{kT} \exp\left(-\frac{Q_0}{kT} \right) \tag{3.10}$$

(ii) If, after going over the saddle point, the dislocation is unstable and slips rapidly over a distance Λ until it is blocked in front of another barrier, it cannot jump back and we have then:

$$v = v^+ = v_0 \exp\left[-\frac{Q_0 - \sigma b \Delta A}{kT} \right] = v_0 \exp\left(-\frac{\Delta G}{kT} \right) \tag{3.11}$$

Let l be the length of the segment of dislocation freed after each successful jump, the elementary strain after each successful jump is lb^2/V in the first case and $l\Lambda b/V$ in the second one; V is the volume of the crystal (see § 2.3.4).

The strain-rate can be written:

$$\dot\varepsilon = Nlb^2 v \tag{3.12}$$

for case (i) with v given by (3.9) and

$$\dot\varepsilon = Nl\Lambda bv \tag{3.13}$$

for case (ii) with v given by (3.11).

N is the number of flow-units (or activation sites) per unit volume.

3.2.2   The force–distance diagram

The quantities involved in the thermally activated overcoming of obstacles are conveniently visualized on the 'force–distance' diagram (fig. 3.6). F (in ordinates) is the force exerted on the dislocation segment of length l by the applied stress:

$$F = \sigma b l$$

The reaction coordinate (in abscissae) is the distance travelled by the dislocation during the activation process. The energy barrier is represented as a hill, of height $\sigma_B bl$, rising over a plain of altitude $\sigma_i bl$; σ_i is the average value of the spatially fluctuating long-range *internal stress*, caused by the other dislocations. The wavelength of the internal stress is large and the maxima cannot be overcome with the help of thermal agitation: a part of the applied stress must therefore be used to move the dislocation against the internal stress; the rest, or *effective stress*, σ_{eff}, helps in overcoming the obstacle.

We can write:

$$\sigma = \sigma_i + \sigma_{eff} \tag{3.14}$$

The dislocation is pushed by the applied stress up the slope of the hill at a height $\sigma_{eff}lb$ above the plain; the remaining height to the hill top can be ascended only with the help of thermal agitation.

The diagram corresponds to a fixed value of strain-rate $\dot{\varepsilon}$ and temperature T.

   (i) At $T = 0$ the obstacle can be overcome only if the applied stress is at least equal to $\sigma_i + \sigma_B$.

  (ii) For $0 > T > T_c$, the obstacle can be overcome with the help of the applied stress and thermal agitation, provided it is not too high or too wide, in which case thermal agitation would be ineffectual and the deformation could only be athermal (fig. 3.7(*a, b*)).

 (iii) For $T > T_c$, T_c being the temperature at which thermal agitation can provide the total energy needed to overcome the barrier $(kT_c > \sigma_B b\Delta A)$, the barrier has disappeared, the dislocation moves athermally under $\sigma = \sigma_i$ (fig. 3.7(*c*)).

We see that *the deformation can be thermally activated only if the thermal energy is of the order of magnitude of the energy barrier*; if it is much lower or

Fig. 3.6. Force–distance diagram for thermally activated overcoming of obstacles ($T =$ const, $\dot{\varepsilon} =$ const). $lb\sigma_B$ height of the barrier, $lb\sigma$ force provided by applied stress, $\sigma = \sigma_i + \sigma_{eff}$ (internal stress + effective stress). Hatched area: energy provided by σ_{eff}. Stippled area: energy provided by thermal agitation.

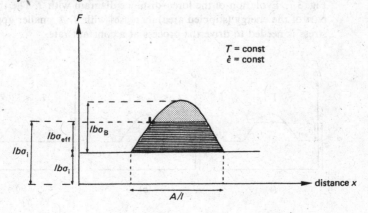

much higher the deformation has to be controlled by some other, athermal process.

In the case of thermally activated deformation, the dislocation can jump over the barrier v times per second, for given $\dot{\varepsilon}$ and T.

- For constant strain-rate conditions, at a given temperature T, the effective stress has just the right value for the obstacles to be overcome at the frequency corresponding to the imposed strain-rate.
- For constant stress conditions, at a given temperature, thermal agitation can cause the dislocations to overcome the obstacles at a definite frequency only, to which corresponds a definite creep-rate.

If the imposed strain-rate is higher, the dislocation must jump successfully over the barrier at a higher frequency v, hence, for a given temperature, the part of the energy to be thermally provided must be smaller, which can be achieved only at the price of a higher effective stress (hence applied stress) (fig. 3.8). *A thermally activated process is strain-rate-sensitive and an increase in strain-rate produces the same effect as a decrease in temperature.*

3.2.3   Activation quantities

To make the step from the absolute reaction-rate analysis, based on the consideration of one activation event, to the thermodynamic macroscopic analysis, we must make the assumption that we can average over the crystal the characteristics of all the flow-units, we will therefore deal with average quantities. We must then define the system and choose the variables.

Fig. 3.7. Evolution of the force–distance diagram with T. The thermal part of the energy (stippled area) increases with T. A smaller applied stress is needed to drive the process at a constant rate.

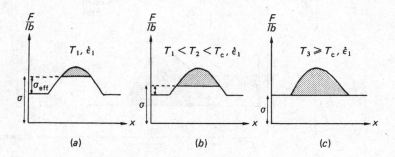

(a)   (b)   (c)

(i) The choice of the system depends on how the stress variable is chosen (Hirth & Nix, 1969):

- If we take the applied stress σ as stress variable, the system has to be the whole crystal, but σ_i must be known.
- If we take instead the effective stress $\sigma_{eff} = \sigma - \sigma_i$, the closed system considered is local: a small volume, about the size of the dislocation segment, around the activation site.

In what follows, we will choose the applied stress as a variable.

(ii) Once the system is chosen, we must express the free energy of activation ΔG in terms of the chosen variables and use that expression to write the strain-rate (see Evans & Rawlings, 1969, and Hirth & Nix, 1969). We will consider here the case where eqs. (3.11) and (3.13) are appropriate. The free energy of activation of the process is equal to the difference of free energy between the final and the initial states and must be provided by thermal agitation. In the initial state (1), the dislocation is in equilibrium under stress in front of the obstacle (on the slope of the hill).

The final state (2) is the activated state in which the dislocation is in the saddle-point position, in thermodynamic equilibrium (assumption required by the absolute reaction-rate theory). Let us consider a reversible and isothermal virtual displacement of the dislocation (along the reaction coordinate x) from the initial to the final state:

$$\Delta G = \int_1^2 [\sigma_B(x, T) + \sigma_i(x, T)]bl\,dx - \sigma b \int_1^2 l\,dx \qquad (3.15)$$

Fig. 3.8. Evolution of the force–distance diagram with strain-rate. For the process to take place at a faster rate (right), at the same temperature, the thermal part of the energy (stippled area) is reduced and the mechanical part is increased, hence the stress increases.

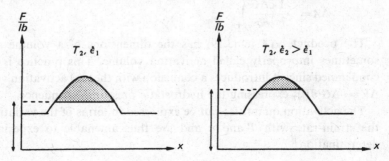

Assuming that l does not depend on x and putting

$$\Delta x = x_2 - x_1 = f(\sigma)$$

we can write:

$$\Delta G = \Delta G_0(T) - \sigma b \Delta A \qquad (3.16)$$

where

$$\Delta G_0(T) = \int_1^2 [\sigma_B(x, T) + \sigma_i(x, T)] bl \, dx$$

$\Delta A(\sigma) = l \Delta x$ is the *activation area*, representative of the extent of the obstacle in the glide plane.

We can write:

$$\Delta G_0(T) = \Delta H_0 - T \Delta S \qquad (3.17)$$

ΔH_0 is the activation enthalpy (for zero applied stress) and ΔS is the activation entropy. The most important source of entropy is the variation of the shear modulus μ with temperature (σ_B and σ_i are proportional to μ). Using (3.11), (3.13), (3.16) and (3.17), we obtain:

$$\dot{\varepsilon} = Nl\Lambda bv_0 \exp\left(\frac{\Delta S}{k}\right) \exp\left[-\frac{\Delta H_0 - \sigma b \Delta A}{kT}\right] \qquad (3.18)$$

or

$$\dot{\varepsilon} = \dot{\varepsilon}_0 \exp\left[-\frac{\Delta H_0 - \sigma b \Delta A}{kT}\right] \qquad (3.19)$$

The activation quantities are defined in terms of ΔG by:

$$\Delta S = -\frac{\partial \Delta G}{\partial T}\bigg)_\sigma \qquad (3.20)$$

$$\Delta H = \Delta G + T \Delta S = \frac{\partial\left(\dfrac{\Delta G}{T}\right)}{\partial\left(\dfrac{1}{T}\right)} \qquad (3.21)$$

$$\Delta A = -\frac{1}{b}\frac{\partial \Delta G}{\partial \sigma}\bigg)_T \qquad (3.22)$$

The product $b\Delta A$ in (3.19) has the dimensions of a volume and is sometimes improperly called activation volume. This practice is to be condemned since it introduces a confusion with the real activation volume $\Delta V = \partial \Delta G / \partial P)_T$ expressing the hydrostatic pressure dependence.

The activation quantities can be expressed in terms of the variations of the strain-rate with T and σ and are thus amenable to experimental determination.

3.2.4 Experimental determinations

(i) The activation enthalpy ΔH is identical to the 'apparent activation energy' Q (§ 3.2.1):

$$\Delta H \equiv Q = -k \frac{\partial \ln (\dot{\varepsilon}/\dot{\varepsilon}_0)}{\partial (1/T)}\bigg)_\sigma \tag{3.23}$$

It can be determined by measuring the slope of an Arrhenius plot of steady-state creep-rates at various temperatures determined on different specimens or by 'temperature jumps' by changing the temperature from T_1 to T_2 as rapidly as possible and measuring $\dot{\varepsilon}_1$ and $\dot{\varepsilon}_2$ when it comes back to quasi-steady state.

$$\Delta H \approx -k \frac{\Delta \ln \dot{\varepsilon}}{\Delta (1/T)} = -k \frac{\ln (\dot{\varepsilon}_2/\dot{\varepsilon}_1)}{1/T_2 - 1/T_1} \tag{3.24}$$

Note that, in eq. (3.23), k is the Boltzmann constant, hence ΔH is the activation enthalpy per atom and is often expressed in electron volts per atom. It is often convenient to consider the activation enthalpy per mole, expressed in calories or joules; in (3.19) and (3.23) k is then replaced by $R = kN_A$ (N_A = Avogadro's number) which is conveniently expressed in calories/mole: $R \approx 2$ cal/mol (1 eV/atom $\approx 23\,000$ cal/mol).

In many cases, $\dot{\varepsilon}_0$ is supposed to be independent of temperature (e.g. eq. (3.24)), this, however, is not strictly true since $\dot{\varepsilon}_0$ depends on the shear modulus μ, which in turn depends slightly on temperature (§ 1.1.2). This is especially true when $\dot{\varepsilon}$ is fitted by a power-law equation, where the stress dependence is entirely contained in the pre-exponential term; in the case of Dorn's equation:

$$\dot{\varepsilon} = A \left(\frac{\sigma}{\mu}\right)^n \exp\left(-\frac{Q}{RT}\right) \tag{3.25}$$

It can be shown that the apparent activation energy given by (3.23) is in fact:

$$Q_{\text{app}} = Q - nR \frac{T^2}{\mu} \frac{d\mu}{dT} \tag{3.26}$$

Failure to take the dependence of the elastic moduli on T into account may result in a slight curvature of the Arrhenius plot (Barrett *et al.*, 1964).

(ii) The activation area ΔA is given by:

$$\Delta A = \frac{kT}{b} \frac{\partial \ln (\dot{\varepsilon}/\dot{\varepsilon}_0)}{\partial \sigma}\bigg)_T \tag{3.27}$$

If $\dot{\varepsilon}_0$ is assumed to be independent of stress we have

$$\Delta A = \frac{kT}{b\sigma} \frac{\partial \ln \dot{\varepsilon}}{\partial \ln \sigma}\bigg)_T \tag{3.28}$$

or in the power-law case:

$$\Delta A = \frac{kT}{b} \frac{n}{\sigma} \tag{3.29}$$

ΔA can be determined from the plot $\ln \dot{\varepsilon}$ against $\ln \sigma$ or from stress jumps during the same creep experiment (the values found may be different, see §1.2.3). The stress sensitivity of the strain-rate, hence the activation area, can also be determined from stress-relaxation curves (Guiu & Pratt, 1964; Rutter *et al.*, 1978).

In stress relaxation, we transform the elastic strain into plastic strain:

$$\dot{\varepsilon} = -M\dot{\sigma}$$

where M is the effective elastic modulus of the machine-plus-specimen system (§1.1.3).

We have:

$$\frac{\partial \ln(-\dot{\sigma})}{\partial \sigma} = \frac{\partial \ln \dot{\varepsilon}}{\partial \sigma} = \frac{b}{kT} \Delta A \tag{3.30}$$

The activation area is therefore determined from the slope of the $\ln(-\dot{\sigma})-\sigma$ plot (fig. 3.9).

If there is no recovery, the internal stress σ_i does not decrease during relaxation and only the effective stress σ_{eff} decreases, the stress tends towards σ_i for infinite time.

Other methods for obtaining an approximate value of the internal stress are the stress-dip method (Solomon & Nix, 1970) and the strain-dip method (Ahlquist & Nix, 1969).

Fig. 3.9. Determination of the internal stress σ_i and the activation area ΔA (b) from stress-relaxation experiments (a).

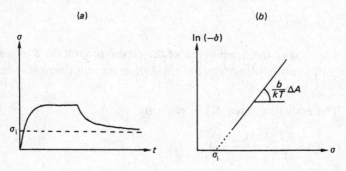

Fig. 3.10. (*a*) Stress-dip method for determining the internal stress σ_i.
(*b*) Strain-dip method for determining the internal stress σ_i.

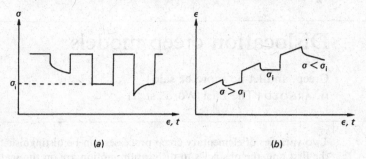

(a)   (b)

In the stress-dip method (fig. 3.10(*a*)), the stress is relaxed by steps until an inverse transient is observed when re-establishing the stress; in the strain-dip method (fig. 3.10(*b*)), for creep tests, the applied stress is lowered by steps during a creep test until a negative creep-rate is observed when re-establishing the stress. This method is somewhat controversial and stress drops have often been found to result only in an incubation period of zero creep-rate (Birch & Wilshire, 1974; Bergman, 1975).

4

Dislocation creep models

Creep, and let no more be said!
M. ARNOLD ('The Last Word', st. 1)

Two varieties of elementary creep processes can be distinguished: in the first one, the obstacles to dislocation motion are on the scale of the dislocation core and are *overcome* by thermal agitation helped by the stress (glide-controlled creep); in the second one, the obstacles are too large to be overcome by thermal agitation but they may *disappear* by diffusion-controlled recovery (recovery-controlled creep) – dislocation motion is not directly thermally activated but it is controlled by a thermally activated process. Models for creep resorting to both types are reviewed. The stress exponent n and the activation energy Q are very weak constraints on the models. Specific problems relative to metallic alloys and ceramics and minerals are discussed: climb dissociation of dislocations, creep of olivine, creep in hydrogen oxide – ice.

4.1   Generalities

The purpose of creep models is to give a physical basis to the empirical constitutive relations. For dislocation creep, the problem is in fact to express the average dislocation velocity in Orowan's equation (§ 2.3.4) in terms of temperature- and stress-dependent processes. The average velocity of dislocations can be written:

$$\bar{v} = \frac{\Delta L}{t_g + t_0} \tag{4.1}$$

where ΔL is the distance the dislocation glides after overcoming an obstacle, t_g is the time taken to glide over ΔL and t_0 is the time taken to overcome the obstacle (assuming that the strain associated with the overcoming is negligible).

Two different cases may be distinguished (table 4.1):

(i) *The obstacles to dislocation motion have dimensions on the scale of the dislocation core* (i.e. a few interatomic distances) – *they can be overcome by thermal agitation helped by an effective stress*, as seen in § 3.2. The dislocation motion is then truly thermally activated and the activation enthalpy is stress-dependent. Two cases can then be envisaged:

- The dislocation glides rapidly after overcoming the obstacle until it meets another one at a distance Λ (eq. (3.13)). We have then in (4.1): $\Delta L = \Lambda$ and $t_g \ll t_0$.

  This situation may occur if the obstacle consists in unravelling an attractive junction at high stresses or if localized obstacles must be overcome by cross-slip of screw dislocation, as in FCC metals at intermediate temperatures.

- The dislocation, after overcoming the obstacle, glides over $\Delta L = b$ and immediately meets the next obstacle; in that case we may speak of *glide-controlled creep* (§ 4.2); the controlling and strain-producing process being glide itself and the obstacles being the Peierls hills, there is no difference between t_g and t_0. The creep kinetics obviously depend much on the dislocation core structure and may be analysed in terms of creation and mobility of double kinks or in terms of spread cores. This situation arises at low or intermediate temperatures in BCC or HCP metals; at high temperatures, the lattice obstacles are too easily overcome by thermal agitation and other processes control the strain-rate with the result that dislocation motion may be indirectly thermally

Table 4.1. *Dislocation creep mechanisms*

OBSTACLES	Distributed on the scale of the core	Localized
can be OVERCOME by thermal agitation with help of σ $\Delta H(\sigma)$	Lattice friction *GLIDE-CONTROLLED CREEP*	Nodes, barriers, attractive junctions Cross-slip may be active
can DISAPPEAR by thermally activated cross-slip $\Delta H(\sigma)$		Mutually locking split screw dislocations Network growth Internal stress *RECOVERY-CONTROLLED CREEP*
can DISAPPEAR by diffusion-controlled climb ΔH		Mutually locking edge dislocations Network growth Internal stress *RECOVERY-CONTROLLED CREEP*

activated or athermal (see below). In minerals, however, where the bonding is ionocovalent or covalent, the Peierls stresses are high, even at elevated temperatures, as evidenced by the frequent straight dislocations along crystallographic directions, and glide-controlled creep may be prevalent (see §4.6).

(ii) *The obstacles to dislocation motion are too wide to be overcome by thermal agitation; however, they may disappear through the operation of a thermally activated process.* This situation arises when the obstacles are created by other dislocations (fluctuating internal stress field or localized tangles or barriers formed by mutually locked-up dislocations). These obstacles can dissolve away by climb-controlled mutual annihilation of the dislocations, climb itself being controlled by diffusion, which in turn is thermally activated; the obstacles disappear and release dislocations at a rate that follows an Arrhenius kinetics. It is, however, important to realize that this process is in fact only indirectly thermally activated as the activated step pertains to the migration of vacancies, not to the motion of dislocations. The activation enthalpy is stress independent. We may speak of *recovery-controlled creep* or, in terms of its stress dependence, of *power-law creep* (see §4.3). In that case, $\Delta L = \Lambda$ and $t_g \ll t_0$, but t_0 is determined by the diffusion-controlled climb velocity of the dislocations.

Cross-slip of screw dislocations towards one another with subsequent mutual annihilation may also be a way for obstacles to disappear, analogous to mutual annihilation of edge dislocations by climb (Poirier, 1976), hence it may be considered as a recovery process, leading to indirect thermal activation of dislocation motion. In this case, however, there is a stress dependence of the activation enthalpy (see §§4.2 and 4.3).

It follows from what we have seen in chapter 3 that, depending on the relative values of the activation enthalpies of the processes, there may exist temperature and stress domains where glide-controlled and recovery-controlled processes act as parallel–concurrent processes. If the creep-rate is really thermally activated (i.e. glide-controlled or cross-slip-controlled) the activation energy is stress dependent; from Orowan's equation and the fact that the dislocation density depends linearly on σ^2 (see §4.3.2), we can expect the creep rate to be of the form:

$$\dot{\varepsilon} = \dot{\varepsilon}_0 \left(\frac{\sigma}{\mu}\right)^2 \exp\left[-\frac{Q(\sigma)}{RT}\right]$$

4.2 Creep by thermally activated dislocation motion

We consider here the processes that must be analysed in terms of stress-assisted thermal overcoming of obstacles (localized or uniformly distributed, as Peierls hills).

4.2.1 Localized obstacles

(i) Cross-slip of screw dislocations can play an important role in the overcoming of obstacles. It is possible to decompose the thermally activated process in the following steps (Friedel, 1964) (fig. 4.1).

A pre-existing constriction on a split screw dislocation (e.g. a constricted intersection jog) can separate into two constrictions; the small region in between the constrictions can split into the cross-slip plane, if it reaches a critical size – it can bow out under the action of the applied stress and become unstable in the cross-slip plane. Escaig (1968*a*) has calculated the elastic energy of the saddle-point configuration and derived an expression for the stress-dependent activation energy of cross-slip in the case of face-centred cubic metals. For high stresses the expression of $Q(\sigma)$ is not analytic but can be numerically computed (fig. 4.2); for low stresses an asymptotic form of the general equation gives a linear dependence in stress for the activation enthalpy:

$$Q = A\frac{\mu b^2}{4\pi}d_0\left[\left(1-\frac{b_0}{d_0}\right)^2 - \alpha\frac{\sigma b}{\gamma_F}\right] \tag{4.2}$$

Fig. 4.1. Cross-slip of a screw dislocation (after Friedel). 1. Pinching of the stacking-fault ribbon. 2. Splitting of a portion into the cross-slip plane. 3. Spreading of a split segment into the cross-slip plane.

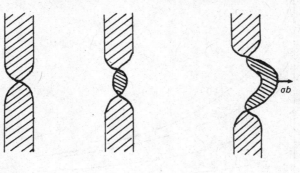

1   2   3

where μ is the shear modulus; b, the Burgers vector; b_0 and d_0 the width of the unsplit and split dislocation respectively; σ, the applied stress; γ_F, the stacking-fault energy and A and α are coefficients.

Cross-slip of screw dislocations is responsible for the stage III with parabolic hardening (following the linear hardening stage II), observed in the stress–strain curves of face-centred cubic metals (Friedel, 1964) and in alkali halides (Davidge & Pratt, 1964; Strunk, 1975) (fig. 4.3). The screw dislocations piled up in front of sessile Lomer–Cottrell locks cross-slip onto other $\{111\}$ planes, or in some cases onto $\{110\}$ and $\{100\}$ planes (Le Hazif & Poirier, 1975; Caillard & Martin, 1982) (fig. 4.4).

The creep of aluminium at intermediate temperatures (0.25 to $0.5T_m$) is likely to be controlled by cross-slip (Dorn & Jaffe, 1961). The work of Myshlyaev (1976) and, above all, the remarkable experiments on creep of thin samples of aluminium inside a transmission electron microscope by Caillard & Martin (1982, 1983) have shown that the controlling steps take place in the dislocation walls formed during the deformation and that cross-slip is the rate-controlling mechanism. The role of cross-slip in the knitting and unravelling of dislocation walls has been theoretically analysed by Friedel (1977).

The analysis of published high-temperature creep data on silicon and germanium (Siethoff, 1983) shows a good agreement with Escaig's (1968a)

Fig. 4.2. Variation of the activation energy for cross-slip with stress. Q and σ are expressed as dimensionless quantities. The figures on the curves are values of $\lambda = d_0/b_0$ (after Escaig, 1968a).

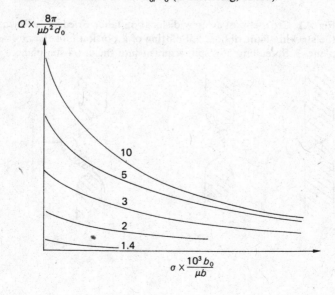

Fig. 4.3. The stress–strain curve of NaCl single crystals. Stage I: easy glide. Stage II: linear hardening. Stage III: cross-slip (for $\sigma > \sigma_{III}$). (After Davidge & Pratt, 1964.)

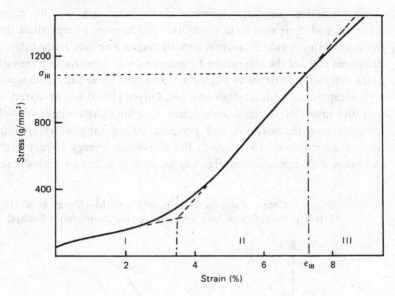

Fig. 4.4. Cross-slip in aluminium. Slip traces in a foil crept *in situ* in HVEM at 520 °C. TEM at 100 keV (courtesy D. Caillard).

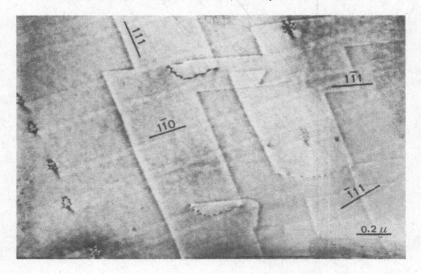

model (fig. 4.5) and strongly suggests that high-temperature creep in these covalent materials is cross-slip-controlled.

(ii) Attractive junctions between dislocations are usually extremely resistant and their destruction requires much more energy than thermal agitation can provide in normal conditions; as a consequence other easier processes control the creep-rate. In some cases, however, creep cannot take place without the attractive junctions being destroyed and this can occur at high temperatures under high stresses. Guyot (1968) has analysed in this way the creep of sintered aluminium powder (SAP) where particles of alumina trap dislocations and produce strong tangles which must be unravelled for creep to proceed; the activation energy is very high and increases with temperature (fig. 4.6) (in fact, it increases with decreasing

Fig. 4.5. Creep of silicon. Data points from Myshlyaev *et al.* (1969) fitted to Escaig's (1968*a*) cross-slip control theory (after Siethoff, 1983).

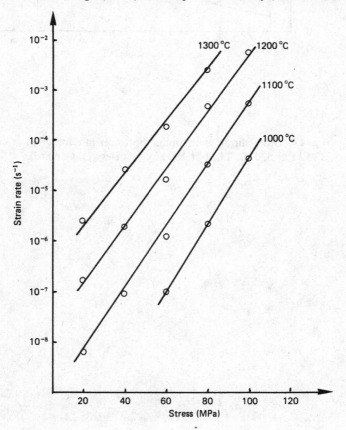

stress, as higher-temperature experiments are conducted at lower stresses). A theoretical calculation of the activation energy for destroying the junction is in good agreement with the experimental observations.

4.2.2 Glide-controlled creep

(i) Dislocation glide controlled by the thermally activated overcoming of the Peierls hills has been essentially studied in body-centred cubic metals at low temperatures, for constant strain-rate experiments. The macroscopic elastic limit corresponds to the temperature-independent stress at which straight and sessile screw dislocations begin to move (Escaig, 1967). The variation with temperature of the macroscopic elastic limit was interpreted in two equivalent ways, either with the double-kink formalism

Fig. 4.6. Activation energy as a function of temperature for SAP (sintered aluminium powder) containing 3.6, 8.3, 11.7% Al_2O_3 particles (after Guyot, 1968).

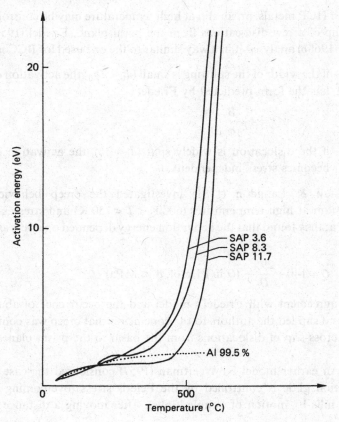

(Guyot & Dorn, 1967) or with the following model of the core structure (see Vitek, 1974): the core of the screw dislocation is seen as simultaneously spread on several potential glide planes, in zone with the dislocation (Kroupa & Vitek, 1964); the stacking-fault ribbons prevent glide in planes other than their own, the result being that the dislocation is blocked unless a high enough stress in conjunction with thermal agitation constricts locally the dislocation and allows the formation of a double kink (Escaig, 1968b). This can be seen as 'continuous' cross-slip, the glide on each plane being limited to the distance to the next trough and the whole process starting anew on the same, or possibly a different, plane (this is the origin of apparently non-crystallographic 'pencil glide').

In BCC metals, this process is mostly effective at rather low temperatures but it may operate as a high-temperature creep process in hexagonal close-packed metals and may be important in some minerals, where the range of processes limited to low temperatures in metals may be considerably extended upwards.

(ii) In HCP metals, prism slip at high temperature may be controlled by cross-slip of screw dislocations from the basal plane; Friedel (1959) and Escaig (1968b) analysed it in a way similar to the one used for BCC metals:

– if the width of the splitting is small ($d_0 < 2b_0$) the activation energy has the form predicted by Friedel

$$Q = A + \frac{B}{\sigma} \tag{4.3}$$

– if the dislocation is widely split ($d_0 > 3b$), the activation energy becomes stress-independent.

Vagarali & Langdon (1981) investigated the creep behaviour of magnesium at high temperatures (600 K $< T <$ 750 K) and stresses above 2.5 MPa; they found that the activation energy depended on stress as in eq. (4.3):

$$Q = 140 + \frac{295}{Q} \quad (Q \text{ in kJ/mol, } \sigma \text{ in MPa})$$

The agreement with Friedel's model and the occurrence of abundant non-basal slip led the authors to the conclusion that creep was controlled by the cross-slip of dislocations from the basal to the prism planes.

(iii) An earlier model by Weertman (1957) considered the case where dislocation glide is controlled by the Peierls stress (overcoming of the Peierls hills by motion of double kinks). After moving a distance L, the dislocations annihilate by climbing to dislocations of opposite sign created on neighbouring slip planes but the temperature is high enough for climb to be easier than glide ($t_0 \ll t_g$ in eq. (4.1)); we have therefore glide-controlled creep.

The dislocation velocity is then given by the absolute reaction-rate theory as:

$$v_g = vL\frac{a}{b}\exp\left[-\frac{Q(1-\pi\sigma/2\tau)}{kT}\right]$$

(4.4)

where v is the frequency of vibration of the dislocation segments, a the distance between Peierls hills and τ the Peierls stress. The creep-rate is calculated by Orowan's formula $\dot{\varepsilon} = \rho v_g b$ where the dislocation density ρ is equal to the product of the density of sources M by the number of dislocations in a pile-up of length L under the applied stress σ ($\propto bL$) and by the average length of a loop ($\propto L$). The velocity is also proportional to L. With the relation

$$L^2 = \frac{2\sigma}{\mu bM}$$

(4.5)

we obtain:

▶ $$\dot{\varepsilon} = 12vaM^{-0.5}\left(\frac{\sigma}{\mu}\right)^{2.5}\exp\left[-\frac{Q(1-\pi\sigma/2\tau)}{kT}\right]$$

(4.6)

At low stresses the creep-rate follows a power law with $n = 2.5$ (assuming that M is stress-independent) and at higher stresses the exponential dependence on σ comes in, giving a curved log $\dot{\varepsilon}$–log σ plot (see § 3.1.3). Weertman derived this model to explain the high-temperature behaviour of creep in zinc where the activation energy is higher than the self-diffusion activation energy, but he pointed out that this model could be useful in the case of non-metallic crystals with high Peierls stresses.

4.3 Power-law creep or recovery-controlled creep

4.3.1 Bailey–Orowan equation

We are concerned here with real power-law creep, i.e. with quasi-steady-state creep, with a constitutive equation of the form $\dot{\varepsilon} \propto \sigma^n$, where the stress exponent n has a physical meaning and does not simply arise from the fit of a straight line through data points on a log $\dot{\varepsilon}$–log σ plot. Power-law creep is recovery-controlled and is only indirectly thermally activated through the processes controlling recovery, of which the principal is diffusion. At high temperatures, the obstacles that can be overcome by thermally activated motion of dislocations eventually disappear (see §§ 3.2.2 and 4.1) i.e. σ_B the height of the barrier becomes equal to zero, since thermal agitation provides the totality of the energy needed to overcome the barrier. The only obstacles that remain to dislocation motion are the long-period, smooth undulations of the internal stress field and these have to be overcome by the applied stress alone. We will use the activation formalism once more to show that it leads in this case to the power-law formulation. Let us write again the expression of the activation free energy:

$$\Delta G = \int_1^2 [\sigma_B(x, T) + \sigma_i(x, T)]bl\, dx - \sigma b \int_1^2 l\, dx \tag{3.15}$$

Taking $\sigma_B \equiv 0$, we have:

$$\Delta G = \int_1^2 (\sigma_i - \sigma)bl\, dx \approx (\sigma_i - \sigma)b\Delta A \tag{4.7}$$

where ΔA is the activation area. Hence:

$$\dot{\varepsilon} = \dot{\varepsilon}_0 \exp\left[\frac{(\sigma - \sigma_i)b\Delta A}{kT}\right] \tag{4.8}$$

Now, the well-known fact that crystals harden when strained and soften when annealed at high temperatures can be translated in terms of internal stress: the internal stress increases with strain and decreases as the annealing time t increases.

After a time dt at high temperature during which the strain produced is d, the internal stress varies by:

▶ $$\Delta\sigma_i = h\, d\varepsilon - r\, dt \tag{4.9}$$

where h and r are coefficients, which we will *not* write as partial derivatives of stress with respect to strain and time, an objectionable formulation since strain and time are not independent variables (McCartney, 1976); we will simply call h the strain-hardening rate and r the recovery rate.

For constant applied stress, (4.8) can be written:

$$\dot{\varepsilon} = \dot{\varepsilon}_0 \exp\left[-\frac{(h\varepsilon - rt)b\Delta A}{kT}\right] \tag{4.10}$$

where $\dot{\varepsilon}$ is the strain rate at time t and $\dot{\varepsilon}_0$ the strain rate at $t = 0$.

Equation (4.10) can be integrated from $t = 0$ to t (McLean, 1966) and yields:

▶ $$\varepsilon = \frac{kT}{\Delta A\, bh} \ln\left(\frac{\dot{\varepsilon}_0 h}{r}\right) + \frac{r}{h}t \tag{4.11}$$

The first term of the right-hand side of eq. (4.11) describes the primary part of the creep curve (compare eq. (1.33) for Andrade creep) and the second term, the steady-state part with a creep rate:

▶ $$\dot{\varepsilon}_s = \frac{r}{h} \qquad (4.12)$$

Eq. (4.12) is known as the Bailey–Orowan equation.

From (4.12) and (4.9), we see that the internal stress σ_i remains constant during steady-state creep, and from eqs. (4.8) and (4.10) that the applied stress has to be just equal to the internal stress for creep to proceed at constant rate. In other terms:

$$\sigma - \sigma_i \equiv \sigma_{eff} = 0 \qquad (4.13)$$

and the exponential dependence in stress disappears. As we will see later, the stress dependence of the recovery rate can take a power-law form; as to the temperature dependence, it is also contained in the recovery rate r, which is often diffusion-controlled. Mitra & McLean (1966) have verified the validity of eq. (4.12) by measuring the values of r and h:

(i) r was determined by suddenly lowering the applied stress by $\Delta\sigma$ during steady-state creep; a zero-creep period followed during the time Δt necessary for σ_i to decrease to adjust itself to the new stress (fig. 4.7). The experiment was done for several values of $\Delta\sigma$ and extrapolated to $\Delta\sigma = 0$:

$$r = \lim_{\Delta\sigma \to 0} \left(\frac{\Delta\sigma}{\Delta t} \right) \qquad (4.14)$$

Note that the shape of the curve after a stress drop is a long-standing object of controversy (see Takeuchi & Argon, 1976a). However, it seems that the existence of zero creep is compatible with an athermal recovery control

Fig. 4.7. Zero creep following stress drop. The recovery rate is taken as the limit of $\Delta\sigma/\Delta t$ for $\Delta\sigma \to 0$ (after Mitra & McLean, 1966).

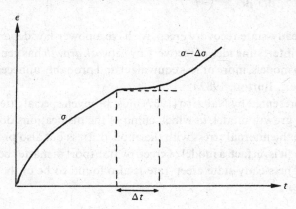

mechanism (Poirier, 1977) and reliable experiments seem to confirm it (Bergman, 1975).

(ii) *h* was determined as the strain-hardening coefficient on a stress–strain curve after creep, for the same value of σ/μ as in the creep experiment.

The agreement in the case of nickel and aluminium is quite good. It is obvious that in the case of recovery-controlled creep the thermal activation formalism is not appropriate. If, nevertheless, the creep-rate were expressed as in eq. (3.19), the activation area ΔA should be found to depend on stress as $1/\sigma$, thereby effectively eliminating stress from the exponential. Such a dependence has indeed been found by Balasubramanian & Li (1970) for a great number of metals, alloys, semi-conductors and ionic solids (fig. 4.8).

The stress dependence of the recovery rate can be calculated if the average mesh size *l* of the three-dimensional Frank dislocation network is taken as the factor that governs the internal stress:

$$\sigma_i \propto \frac{\mu b}{l} \tag{4.15}$$

Friedel (1964, p. 239) has shown that the mesh of the network can grow by climb of dislocations, owing to exchange of vacancies between them, so that the total energy decreases; the rate of increase dl/dt of the average mesh size is shown to be inversely proportional to its size, and proportional to the self-diffusion coefficient *D*.

$$\frac{dl}{dt} \approx D\frac{\mu b^3}{lkT} \tag{4.16}$$

It follows (McLean, 1966) that:

$$r = \frac{d\sigma_i}{dt} = \frac{d\sigma_i}{dl} \cdot \frac{dl}{dt} \propto D\left(\frac{\mu b^3}{l}\right) \propto D\sigma_i^3 \tag{4.17}$$

As $\sigma_i = \sigma$ for steady-state recovery creep, we have a power-law dependence with $n = 3$. The interesting idea of recovery by *network growth* has generated a spate of creep models, more or less equivalent and probably unnecessarily sophisticated (e.g. Burton, 1982a).

The model presented by Nabarro (1967) must be given special attention: it is a network growth model, in which climb of the dislocations does not just equilibrate the internal stress with the applied stress but also produces the creep strain (it is, in fact, a model of creep by transport of matter between dislocations). The steady-state creep-rate is also found to be of the form:

$$\dot{\varepsilon} \propto D\sigma^3$$

Fig. 4.8. Activation area for creep of BCC metals and covalent semi-conductors, varies as $1/\sigma$ (after Balasubramanian & Li, 1970).

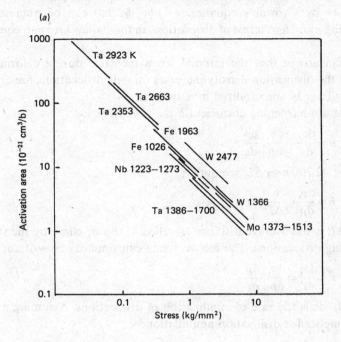

(a)

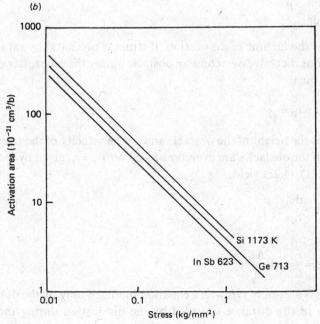

(b)

4.3.2   Equivalence of the Bailey–Orowan and Orowan equations

We will show here that the formulation of quasi-steady-state creep-rate by Orowan's equation $\dot\varepsilon = \rho bv$ (§ 2.3.4) can be obtained by expressing r and h in terms of dislocations in the Bailey–Orowan equation $\dot\varepsilon = r/h$.

We can accept that the internal stress increases during deformation because the dislocation density increases (mobile dislocations are created but may later be immobilized in networks).

The strain-hardening coefficient h may be written:

$$h = \frac{d\sigma_i}{d\varepsilon} = \frac{d\sigma_i}{d\rho} \cdot \frac{d\rho}{d\varepsilon} \tag{4.18}$$

From eq. (2.70), $\varepsilon = \rho\Delta L$, we obtain

$$h = \frac{d\sigma_i}{d\rho} \cdot \frac{1}{b\Delta L} \tag{4.19}$$

where ΔL is the average distance travelled in slip or climb by the strain-producing dislocations. The recovery rate can similarly be written:

$$r = \frac{d\sigma_i}{dt} = \frac{d\sigma_i}{d\rho} \cdot \frac{d\rho}{dt} \tag{4.20}$$

where $d\rho/dt$ is the rate of annihilation of dislocations. Assuming a first-order kinetics for dislocation annihilation:

$$\frac{d\rho}{dt} = \frac{\rho}{\tau} \tag{4.21}$$

where τ is the lifetime of dislocation. If strain is produced by rapid slip of dislocations that have overcome an obstacle during the time t_0 (see eq. (4.1)), we have then:

$$\tau \approx t_0 = \frac{d}{V_0} \tag{4.22}$$

where d is the height of the obstacle and v_0 the velocity of the overcoming process. If the obstacles are overcome by climb, $v_0 = v_c$, given by (2.69). Eqs. (4.20), (4.21), (4.22) yield:

$$r = \frac{d\sigma_i}{d\rho} \rho \frac{v}{d} \tag{4.23}$$

Hence:

▶ $$\dot\varepsilon = \frac{r}{h} = \rho b \frac{\Delta L}{d} v_c \tag{4.24}$$

Eq. (4.24) is a form of Orowan's equation modified only by the dimensionless ratio of the distance travelled by the dislocation during the strain-producing step to the distance travelled during the overcoming of an obstacle.

It is easy to prove that *the 'natural' stress dependence for any climb-controlled recovery creep is a power law with stress exponent* $n = 3$ (Weertman, 1972; Stocker & Ashby, 1973a; Brown & Ashby, 1980). We have only to express the stress dependence of the climb velocity and of the dislocation density. In its linearized form, the climb velocity depends linearly on the stress: $v_c \propto \sigma$. As for the dislocation density, for whatever uniform repartition of dislocations (as, for instance, two- or three-dimensional networks), it must be of the form:

$$\rho \propto \frac{1}{l^2} \tag{4.25}$$

where l is a characteristic length of the dislocation array (e.g. mesh size for a three-dimensional network).

As the stress field of dislocations varies as the inverse of the distance, any dislocation configuration in equilibrium will have a characteristic length:

$$l \propto \sigma_i^{-1} \tag{4.26}$$

where σ_i is the internal stress field. It follows that

$$\rho \propto \sigma_i^2 \tag{4.27}$$

and, if we remember that $\sigma = \sigma_i$, carrying (2.69) and (4.27) into (4.24) yields:

$$\dot{\varepsilon} \propto D\sigma^3$$

A power-2 dependence comes from the dislocation density and a power-1 from the climb velocity. *The physics of the power-law creep boils down to a geometrical relation (4.25), the 1/r dependence of dislocation stress fields and a climb velocity linearized in stress.*

4.3.3 The Weertman model

In the crowd of all the basically equivalent recovery-creep models that clutter the literature, Weertman's (1968, 1972) model stands apart and must be given special attention, because of its importance in the literature as the first physical model of creep, and also because nearly all the other models can be derived from it.

In this model, dislocations are produced by sources in parallel planes, distant by d, and their edge portions are mutually trapped, forming relaxed multipoles (fig. 4.9); when the head dislocations annihilate by climbing toward each other over $d/2$, the blocked loops can then glide over the average distance $L/2$ (L radius of the head loop) and another loop is emitted by the source.

In Orowan's equation, we can write from (4.1)

$$\bar{v} = \frac{L}{d} v_c$$

since $t_0 = v_c d/2 \gg t_g$ and $\Delta L = L/2$, hence:

$$\dot{\varepsilon} = \rho b \frac{L}{d} v_c \tag{4.28}$$

(similar to 4.21).

The active dislocation density is equal to the product of the density of sources M by the number of dipoles per source (taken as equal to $\approx L/6d$) by the average length of a loop: $2\pi \times L/2$:

$$\rho = \frac{\pi M L^2}{6d} \tag{4.29}$$

Assuming that there is only one source per region of radius L and height d, we have

$$M = \frac{1}{\pi d L^2} \tag{4.30}$$

Hence:

$$\rho = \frac{1}{6d^2} \tag{4.31}$$

Carrying (4.31) into (4.28), expressing L by (4.30), and using an expression of the climb velocity analogous to (2.69), we find:

$$\blacktriangleright \quad \dot{\varepsilon} = \alpha \frac{D}{b^{3.5} M^{0.5}} \left(\frac{\sigma}{\mu} \right)^{4.5} \frac{\mu \Omega}{kT} \tag{4.32}$$

Fig. 4.9. Weertman's model. Edge dislocations emitted by sources (S) in parallel planes are mutually trapped and climb towards each other (arrows).

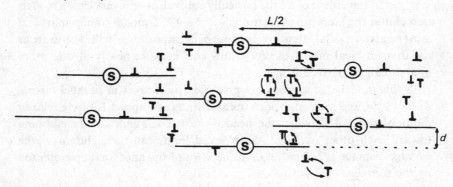

We see that if the density of sources is assumed to be stress-independent, the power law has a stress exponent $n = 4.5$. This is one of the reasons of the popularity of the model, as $n = 4.5$ is often found in creep experiments. With the assumption that $M = \rho/d$, we find $n = 3$.

Weertman (1972) has shown that by making various assumptions on the stress dependence, or absence thereof, of several parameters or by introducing stress concentration factors (especially in the case of pile-ups of dislocations against subgrain boundaries), it is possible – somewhat artificially – to obtain stress exponents as high as 6. Some of the considered '*ad hoc*' assumptions are no doubt valid, especially when their effect is limited to raising n up to 4 or 5. It seems clear, however, that power exponents higher than 5 do not indicate 'real' power-law creep and result from the inadequacy of the log $\dot{\varepsilon}$–log σ plot in those cases.

4.3.4 Power-law creep and self-diffusion

Until quite recently, it was an undisputed belief that diffusion-controlled power-law creep obtained at temperatures higher than about half the absolute melting temperature and for all solids: metals, alloys, ceramics, minerals, etc. (Mukherjee *et al.*, 1969; Weertman, 1972; Takeuchi & Argon, 1976*a*). The rationale for this belief lay in the interpretation of a considerable body of experimental data and an apparently impressive correlation between activation energies for creep and activation energies for diffusion (self-diffusion in the case of metals, diffusion of the slower species in the case of ceramics and minerals, diffusion of the solvent or the solute in alloys); in the rarer cases where activation volumes for creep and diffusion had been measured, they were also found to be similar. Poirier (1978) critically re-examined the source of the data and found that some doubts might be raised as to the general applicability of the statement that all materials creep by diffusion-controlled power-law creep above $0.5 T_m$. In many cases, it was found that Arrhenius plots built from original unedited data were noticeably curved and that the more or less straight portions, with a slope compatible with the self-diffusion activation energy, did not necessarily lie at the highest temperatures (fig. 4.10). The activation volumes for creep and diffusion were sometimes found equal, in temperature domains where the apparent activation energy was not equal to the self-diffusion energy; besides, it can be shown that the activation volume essentially reflects the dependence on pressure of the activation energy through the dependence on pressure of the elastic moduli (see chapter 5), hence the activation volume for creep should be roughly the same for glide- or diffusion-controlled processes and cannot be used as a valid criterion.

Diffusion-controlled creep certainly exists in some temperature range, but it is not the sole high-temperature mechanism and whether it is active or not in a given temperature domain depends on the relative values of the activation energies for diffusion and for the competing process. Cross-slip is a potential competitor for diffusion, as it can act as a parallel–concurrent process along with climb of edge segments; it may for instance control the creep of hexagonal metals above $0.7T_m$ and of copper between 0.5 and $0.7T_m$. The view that high-temperature creep might be controlled by cross-slip has met with a strong opposition (Sherby & Weertman, 1979). The main argument put forward is the following: if the edge segments climb and if the screw segments do not cross-slip because it is too difficult, the kinks that will necessarily connect the edge and screw segments are able to move conservatively down the screw segments, which are forced to move up through the motion of these kinks without cross-slipping (?); climb is thus rate-controlling: now, if the screw segments cross-slip easily, the jogs on the edge segments cannot move conservatively and climb is again rate-controlling. The unstated and unquestioned assumptions are evidently:

(i) that a non-conservative diffusion process is necessarily always more difficult than any other, and

Fig. 4.10. Arrhenius plots for several metals. The plots are conveniently displayed by not labelling the axis. Values of Q are given in kcal/mol. The double lines indicate the regions where $Q = Q_{sd}$ (after Poirier, 1978).

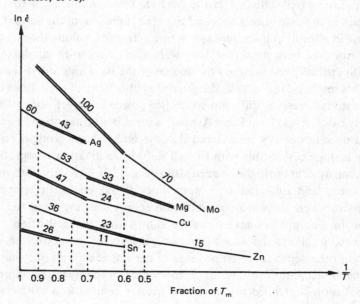

(ii) that a kink always moves easily along a screw dislocation, even if the dislocation is split and cannot cross-slip easily.

We prefer to believe that climb and cross-slip are somehow symmetrical and that cross-slip and climb both acting as parallel–concurrent processes (fig. 4.11) may control the creep-rate in different domains (Poirier, 1976, 1979) (see §4.2). The curvature of the Arrhenius plot is not seen by Sherby & Weertman (1979) as an argument against diffusion-controlled creep in some domains, but as due to the operation of pipe diffusion (with a lower activation energy) at lower temperatures. This is certainly a possibility, although it could not explain the departure from bulk-diffusion activation energy at high temperatures. Spingarn *et al.* (1979) have proposed a creep model where climb is controlled by bulk diffusion and pipe diffusion in parallel. The climb velocity by pipe diffusion is found to be:

$$v_c^d = 12 \frac{D_d b}{l^2} \frac{\sigma b^3}{RT} \tag{4.33}$$

where D_d is the pipe-diffusion coefficient, and l the length of the dislocation segment. Using Orowan's equation

$$\dot{\varepsilon} = \rho b (v_c^{sd} + v_c^d) \tag{4.34}$$

where v_c^{sd} and v_c^d are the climb velocity by bulk and pipe diffusion respectively, the creep-rate can be written:

▶ $$\dot{\varepsilon} = \frac{\mu b}{RT} \left[D_{sd} + 12 \left(\frac{\sigma}{\mu} \right)^2 D_d \right] \left(\frac{\sigma}{\mu} \right)^3 \tag{4.35}$$

Fig. 4.11. Climb and cross-slip as parallel–concurrent processes. Edge portions climb towards each other by migration of jogs (J). Split screw portions (hatched) cross-slip towards each other by migration of kinks (K).

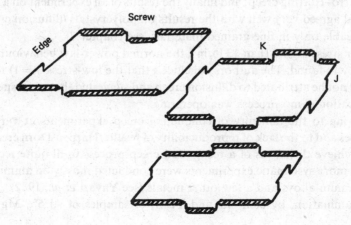

The effective diffusion coefficient depends on the stress through the dislocation density, for in (4.32), $1/l^2 \approx (1/b)^2 (\sigma/\mu)^2$, which introduces a σ^3 dependence for v_c.

This model gives a curved Arrhenius plot, as well as an apparent value of n higher than 3 in a limited stress range. However, the results predicted by the model are not in good agreement with the behaviour of a number of pure metals and the predicted creep-rates are significantly lower than the measured creep-rates.

So far, without discounting a possible effect of pipe diffusion on the curvature of Arrhenius plots, it is difficult not to recognize that diffusion-controlled creep is probably not the only active mechanism at high temperatures. *Let us emphasize again that the determination of the stress exponent n and the activation energy Q in limited domains of stress and temperature do not provide enough information to pinpoint the creep process: n and Q are very weak constraints. In most cases, examination of dislocation configurations by transmission electron microscopy is a must.*

4.3.5   Harper–Dorn creep

Harper & Dorn (1957) found evidence that aluminium poly-crystals deformed in tensile creep at high temperature ($647\,°C = 0.99 T_m$) exhibited a Newtonian viscous behaviour at very low stresses: the Zener–Hollomon parameter varied linearly with stress (fig. 4.12) for stresses lower than $13\,\text{lb/in}^2$ ($0.09\,\text{MPa} \approx 3.3 \times 10^{-6}\,\mu$). Although this behaviour is usually associated with Nabarro–Herring diffusion creep (see chapter 7), Harper & Dorn found that their results were incompatible with it. In particular, they observed primary creep and recovery following stress drop, phenomena that do not exist in Nabarro–Herring creep; also the observed creep rates were three orders of magnitude greater than those calculated for Nabarro–Herring creep; and finally the results of an experiment on a single crystal agreed very well with the results on polycrystals (diffusion creep is noticeable only in fine-grained materials, see chapter 7).

For stress higher than $13\,\text{lb/in}^2$, the normal power-law behaviour, with $n \approx 4$, was found. The authors concluded that the low-stress ($n = 1$) regime could not be attributed to diffusion creep and suggested that an unspecified dislocation climb process was operating.

Owing to the difficulty of conducting creep experiments at very low stresses and to the lack of reproducibility of results, Harper–Dorn creep did not achieve the status of a recognized creep process until quite recently, when more systematic experiments were conducted mostly on aluminium, aluminium alloys and a few other metals (see Yavari *et al.*, 1982).

Examination, by etch-pits and TEM, of samples of Al 5% Mg alloy deformed in the Harper–Dorn creep regime revealed that the dislocation density was quite low ($\rho \approx 5 \times 10^3/\text{cm}^2$) and independent of stress and that dislocation segments seen in TEM were mostly of edge character, no subgrains being seen (Yavari *et al.*, 1982). These results led Langdon & Yavari (1982) to propose that Harper–Dorn creep arises from the climb of edge dislocations under saturated conditions (i.e. when the concentration of jogs is high enough for climb to be controlled by vacancy diffusion to and from the dislocations; this is assumed in most climb-controlled-creep models). Although this is not explicitly stated, it seems that Langdon & Yavari consider that strain is produced by climb of the edge dislocations; their model therefore is equivalent to Nabarro's (1967) with constant dislocation density, i.e. to diffusion creep by exchange of vacancies between dislocations rather than grain boundaries. Mohamed & Ginter (1982) also investigated Harper–Dorn creep in aluminium but achieved large creep strain ($\varepsilon \approx 20\%$) in quasi-steady-state, which supports the contention that Harper–Dorn creep results from a genuine creep process and is not just a transient. They suggest that the dislocation multiplication necessary to obtain large strains proceeds by a cross-slip mechanism.

Fig. 4.12. Zener–Hollomon parameter as a function of stress for aluminium. Note the linear variation of Z with σ (slope on the log–log plot) at low stresses (after Harper & Dorn, 1957).

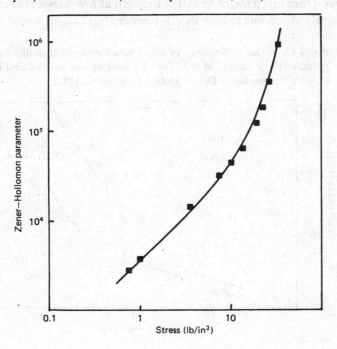

Harper–Dorn creep has also been found in non-metallic materials. Coath & Wilshire (1977) and Dixon-Stubbs & Wilshire (1982) investigated the creep behaviour of lime, CaO, at $T = 1473$ K; for both polycrystals and single crystals they found a low-stress regime with $n = 1$ and a high-stress regime with $n = 5$ (fig. 4.13). For single crystals, the Newtonian creep appears at stresses lower than 20 MPa (or about $2 \times 10^{-4} \mu$); as in the experiments on aluminium, the creep curve exhibits a primary stage and zero-creep recovery periods follow stress drops. The authors reasonably conclude that there is no fundamental difference in the processes occurring at high and low stresses; they suggest that the low-stress behaviour may be accounted for by a recovery-controlled network growth model (§ 4.3.1) with stress-independent dislocation density. The same behaviour is exhibited by single crystalline $KZnF_3$, a fluoride with cubic perovskite structure (Poirier *et al.*, 1983), which crept at low stresses ($5 \times 10^{-5} \mu \leqslant \sigma \leqslant 2.3 \times 10^{-4} \mu$) and high temperature ($T \geqslant 0.9 T_m$).

It seems therefore quite reasonable to account for Harper–Dorn creep by the operation of the same processes as control power-law creep at higher stresses, the difference being that the dislocation density is stress-dependent at high stresses and stress-independent at low stresses. Simple inspection of Orowan's equation (4.24) shows that the only remaining stress dependence comes from the climb velocity and is therefore linear.

Harper–Dorn creep might be rather frequent at low stresses in ceramics and minerals where, due to high Peierls stresses, dislocations would tend to

Fig. 4.13. Stress dependence of strain-rate for single crystalline and polycrystalline lime (CaO), at 1473 K; $n = 1$ at low stresses and $n = 5$ at high stresses (after Dixon-Stubbs & Wilshire, 1982).

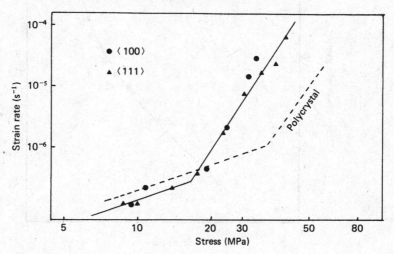

be quite straight and little responsive to stress increases as long as the stresses remain low ($\sigma/\mu \approx 10^{-5}$–$10^{-4}$). However, dislocations have to multiply to allow large strains (up to 30% in $KZnF_3$), and, whatever multiplication process operates, it must be stress-insensitive at low stresses.

4.4 Creep of solid-solution alloys

4.4.1 Empirical classification

Sherby & Burke (1967) and Cannon & Sherby (1970*a*) proposed to divide solid-solution metallic alloys into two categories, on the basis of certain characteristic behaviours.

(i) Class I alloys, exhibiting:
- a power-law dependence in stress with $n \approx 3$,
- little or no primary creep,
- no dependence on the stacking-fault energy γ_F,
- little or no formation of subgrains (polygonization) during creep.

(ii) Class II alloys, exhibiting:
- a power-law dependence in stress with $n \approx 5$,
- normal primary creep,
- a dependence on the stacking-fault energy,
- creep polygonization.

Class II alloys were therefore thought to deform by the same process as pure metals, i.e. climb-controlled power-law creep, the dislocations gliding without hindrance from the solute atoms. Class I alloys were thought to deform by a glide-controlled process in which interaction between solute atoms and dislocations leads to a linear viscous motion of dislocations (see §4.4.3).

Using a compilation of experimental results on various alloys, these authors proposed that alloys could be assigned a category on the basis of the values of two key parameters: the elastic Young's modulus of the solvent, E, and the size difference between solute and solvent atoms, e. Alloys having a large Young's modulus would belong to class II; alloys for which e is large would belong to class I (strong interaction between dislocations and solute atoms). These criteria, however imprecise, were nevertheless found to break down in certain cases; besides, the stress level and the solute concentration should obviously have an influence in determining which behaviour an alloy would exhibit.

Mohamed & Langdon (1974) established a more sophisticated criterion by equating empirical expressions for climb-controlled and viscous glide-controlled creep rate. (Their classes I and II are respectively identical to

Sherby & Burke's classes II and I.) In their derivation, they used an empirical correlation between climb-controlled creep-rate and stacking-fault energy (fig. 4.14)

$$\dot{\varepsilon} \propto \left(\frac{\gamma_F}{\mu b}\right)^3 \tag{4.36}$$

and they assumed that the diffusion coefficient for solute atoms was roughly equal to the weighted diffusion coefficient proposed after Herring, by Weertman (1968) for climb in alloys:

$$D_{HW} = \frac{D_A^* D_B^*}{N_A D_B^* + N_B D_A^*} \tag{4.37}$$

Fig. 4.14. Empirical correlation between creep-rate and stacking-fault energy. Creep-rate, stacking-fault energy and the constant applied stress are expressed as dimensionless quantities (after Mohamed & Langdon, 1974).

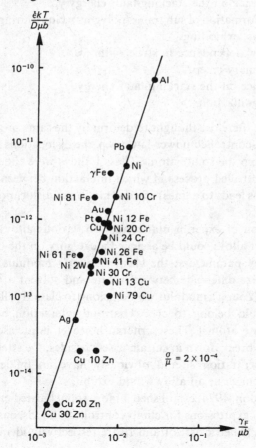

where D_A^*, D_B^* and N_A, N_B are respectively the tracer diffusion coefficient of constituents A and B of the binary AB alloy and the atomic fractions of A and B.

Using experimental results on an alloy Al 3% Mg to fix the adjustable parameters, Mohamed & Langdon find that the criterion for viscous glide is:

$$\frac{8 \times 10^{12} \sigma^2}{k^2(1-v)} \left(\frac{\gamma_F}{\mu b}\right)^3 > \frac{T^2}{e^2 c b^6} \tag{4.38}$$

where k is Boltzmann constant, v Poisson's ratio, c the solute concentration and all the other symbols have been defined above. According to this relation, viscous glide is favoured over climb-controlled creep at high stresses and lower temperatures and for large stacking-fault energies, large size factor of solute and high solute concentration.

Cannon & Sherby (1970b) proposed to extend the concept of two classes to solid-solution 'alloys' of ionic compounds; creep experiments on NaCl–KCl solid solution led them to conclude that they behaved as class I alloys, a solute atmosphere of Na^+ and K^+ ions, in a ratio different from the average composition, diffusing along with the dislocations during glide-controlled creep.

4.4.2   Viscous-glide creep models

Weertman (1957) first proposed a creep model in which dislocations glide with a linear viscous dependence on stress of their velocity. In steady-state creep, dislocations in pile-ups move ahead when the head dislocations annihilate by climb; another dislocation is then created at the source. As in the case of glide controlled by Peierls' stress (§ 4.3.2), climb is easier than glide. The average velocity of dislocations in the pile-up is shown to be proportional to $\sigma b n$, where n, the number of dislocations in the pile-up, is given by $n \propto bL$ (L is the length of the pile-up). The average glide velocity is therefore:

$$v_g = A\sigma^2 b^2 L$$

where A depends on the microcreep viscous mechanism. For a constant density of sources M (related to L by 4.5) and an average length of dislocation proportional to L, the creep rate is:

$$\dot{\varepsilon} \propto AL^2 M\sigma^2 b^2 \propto Ab^2\sigma^3 \tag{4.39}$$

Weertman's theory does not take into account the details of the viscous-glide mechanism, it only uses the assumption that it is linear, i.e. the velocity of one dislocation is proportional to the stress (the apparent dependence of v_g on σ^2 comes from the fact that the number of dislocations in a pile-up is proportional to σ, and the effective stress on a dislocation is $\sigma n \propto \sigma^2$).

Several viscous drag processes can be envisaged:

- Segregation of solute atoms in an atmosphere around the moving dislocations (Cottrell's atmosphere) that must diffuse to follow the dislocation.
- Chemical interaction of solute atoms with the different structure of a stacking fault resulting in segregation of solute atoms on the fault ribbon of extended dislocations (Suzuki effect).
- Destruction of short-range order through the motion of dislocation (Fisher's mechanism).
- Local ordering of solute atoms in the stress field of the dislocation (Snoek mechanism), etc.

More recently, models more elaborate than Weertman's have been proposed for glide-controlled climb involving drag by Cottrell's atmospheres (Takeuchi & Argon, 1976*b*; Burton, 1982*b*). Despite their sophistication (or possibly because of it and of the great number of assumptions they rely on) it is not clear that they contribute much to the physical understanding of creep in solid solutions.

Takeuchi & Argon (1979) have investigated in more detail the physical process of viscous drag of a Cottrell atmosphere on glide and climb by computer simulation.

4.4.3   Discussion: stacking-fault energy and diffusion coefficient

Understanding the physical processes of creep in solid-solution alloys is a formidable task because it compounds most of the difficult problems in physical metallurgy, especially if the alloys are concentrated; let us list a few:

- What is the structure of the alloy at the atomic scale? Even though at high temperature the alloy is in the solid-solution domain, the phase diagram is usually complicated at low temperatures and this may result in short-range order or clustering in certain composition ranges, with momentous consequences on the dislocation motion.
- What is the specific mode of interaction of solute atoms with dislocations: Cottrell atmospheres, segregation on the line or on stacking faults made wider by the segregation itself?
- What diffusion coefficients must be used? Even in dilute alloys there are five different jump frequencies for a vacancy instead of one in pure metals (W_0 – exchange of a vacancy with a solvent atom far from the solute; W_1 – vacancy turning around a solute atom;

W_2 – exchange of a vacancy with a solute atom; W_3 – jump of a vacancy removing it from the vicinity of a solute atom; W_4 – jump of a vacancy bringing it in the vicinity of a solute atom), the number of jump frequencies increases with the clustering of solute atoms (pairs, triplets, etc.) and the atomistic problem of diffusion is intractable in concentrated alloy.

Before discussing in somewhat more detail some of these problems, it is necessary to strike a note of caution as to the interest or validity of general models of creep of alloys resting on empirical compilation of stress exponents and activation energies of creep for various alloys. It is my personal opinion (hence subject to dissent) that the concept of two classes of alloys, although possibly comfortable, is a definite hindrance in the quest for physical processes.

As we have already seen, the stress exponent n has little physical significance *per se* and cannot be used to discriminate between mechanisms, even in the case of pure metals. Now, it is certainly true that in some alloys glide is easier than climb and climb is easier than glide in others, but there is no reason to believe that this is reflected by a change from 5 to 3 in the value of n.

The case of concentrated alloys is especially worrisome because of the complexities of the low-temperature phase diagrams. What are we to make of the fact that in gold–nickel alloys n decreases from 5.6 in pure Au and Ni to about 3 for intermediate compositions (fig. 4.15), when we know that there is a miscibility gap in the phase diagram at lower temperature? What

Fig. 4.15. Variation of stress exponent with composition for gold–nickel solid-solution alloys (after Sellars & Quarrell, 1961).

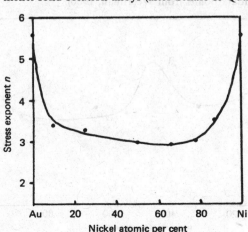

about the variation of n across the concentration range from 0 to 1, in lead–indium alloys (fig. 4.16)?

All criteria for viscous glide (e.g. eq. (4.38)) are based on some microscopic model, like the drag of a solute atmosphere, which has physical meaning only for dilute alloys. Indeed, what is the solute and what is the solvent in a 50 at.% solid-solution alloy? What is an atmosphere around a dislocation? If we take $c = 1$ in eq. (4.38), creep should be even more glide-controlled than with $c = 10^{-2}$ despite the fact that we are dealing with 'pure solute', but it is indicated nowhere in the model that it is to be restricted to dilute alloys, and the compilation on which the alloy classes concept rests includes concentrated alloys (see Table I in Cannon & Sherby, 1970a).

Two parameters are especially worth discussing: the stacking-fault energy and the diffusion coefficient.

(i) The idea that the stacking-fault energy γ_F is a relevant parameter that should be introduced in the creep-rate expression was first put forward by Barrett & Sherby (1965) for pure metals. Even though the correlation between $\dot{\varepsilon}$ and γ_F seems satisfying (fig. 4.14), it is fraught with difficulties. Firstly, γ_F is not an independent variable like T or σ; different values of γ_F correspond to different pure metals, each with its own dislocation core structures. Secondly, the observation that in many alloys (e.g. CuAl) the width of the split dislocations considerably increases with the solute concentration (in the solid-solution range) has often been construed as

Fig. 4.16. Variation of stress exponent with composition for lead–indium alloys (data points from Weertman, 1960).

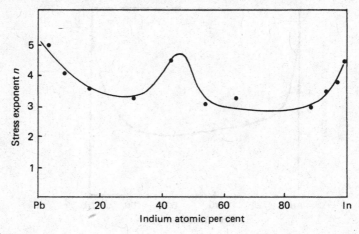

meaning that, for instance, aluminium lowers the stacking-fault energy of copper. This formulation is not correct, since, in adding aluminium atoms to copper we produce an alloy in which the solute atoms segregate to the stacking faults (Suzuki effect) with the result that the width of the equilibrium split dislocations increases (see Hirth, 1970, for a thermo-dynamic treatment); once again, the stacking-fault energy of copper has not been reduced – we have changed copper into an alloy with a lower value of γ_F. Moreover, the values of γ_F are most of the time determined by measuring the width of dislocations or extended triple nodes in TEM at room temperature, but at high temperatures where creep occurs the equilibrium concentration of solute atoms at the stacking fault decreases (solute evaporation) and the width of the dislocation accordingly decreases (hence the apparent value of γ_F increases). It is quite possible that there is an effect of the dislocation width on creep, indeed it is to be expected that kink and jog formation on extended dislocations should be more difficult, with obvious consequences on glide and climb (Burton, 1982c; Argon & Moffatt, 1981), but *a correlation between high-temperature creep rate and room-temperature stacking-fault energy of alloys is meaningless and its existence, at best, fortuitous*: Orlova *et al.* (1971) have found no correlation in the case of Cu–Al alloys up to 16 at.% Al. Finally, jog and kink migration on an extended dislocation is presumably different in pure metals and in alloys with solute segregation on the fault.

(ii) In diffusion-controlled creep models for pure metals, the self-diffusion coefficient appears in the expression of the climb velocity by the product $N_v D_v$ (eq. (2.13)) where N_v and D_v are respectively the atomic fraction and the diffusion coefficients of vacancies. For a dilute solid solution, it is still appropriate to use the self-diffusion coefficient of the solvent. For a concentrated alloy AB where the atomic fractions of constituents A and B are comparable, the notion of solute and solvent disappears, vacancies are not identifiable as vacancies of A or B (as in ionic crystals with distinct sublattices), they are simply vacant sites into which atoms of A and B jump with different jump frequencies. The diffusion coefficient to be used in any diffusion-controlled creep model of alloys must be a mean diffusion coefficient $\bar{D} = N_v D_v$; it should be expressed in terms of the tracer diffusion coefficients of A and B in AB, D_A^* and D_B^*, which can be experimentally determined. Brebec & Poirier (1975), in a general thermo-dynamical treatment, have considered two extreme cases:

– The alloy remains homogeneous, i.e. there are no concentration gradients of A and B, then:

$$\bar{D} = \frac{\varphi}{f} (N_A D_A^* + N_B D_B^*) \tag{4.40}$$

$$\varphi = 1 + \frac{\partial \ln \gamma_A}{\partial \ln N_A} = 1 + \frac{\partial \ln \gamma_B}{\partial \ln N_B} \tag{4.41}$$

φ is Darken's thermodynamic factor, where N_A, N_B and γ_A, γ_B are respectively the atomic fractions and the activity coefficients of A and B. The correlation factor f is close to 1.

This case corresponds to a transient regime before steady-state diffusion is established.

– In the steady state, matter is transported at constant composition, but, as the two species A and B diffuse at different rates, concentration gradients are set up tending to slow the faster species and accelerate the slower one; then (if there is no coupling between fluxes of A and B):

$$\frac{1}{\bar{D}} = \frac{1}{\varphi} \left(\frac{N_A}{D_A^*} + \frac{N_B}{D_B^*} \right) \tag{4.42}$$

or

$$\bar{D} = \varphi \frac{D_A^* D_B^*}{N_A D_B + N_B D_A^*} \tag{4.43}$$

This expression is almost identical to the one derived by Weertman (1968) after Herring.

The activation energy of $\bar{D}$ can be calculated as a function of N_A (or N_B); no agreement is found between its variation and that of the activation energy of creep in the case of gold–nickel and copper–nickel alloys.

Finally let us note that there is no justification in using the chemical or interdiffusion coefficient $\bar{D}$ determined by experiments of diffusion on couples of pure metals A and B:

$$\tilde{D} = \varphi(N_B D_A^* + N_A D_B^*) \tag{4.44}$$

$\tilde{D}$ corresponds to diffusion in steep concentration gradients in A and B but with a zero concentration gradient of vacancies, which is inconsistent with diffusion-controlled creep in a homogeneous alloy. (Note that (4.44) is different from (4.40).) Besides, $\tilde{D}$ does not tend towards the self-diffusion coefficient of pure metals if N_A or N_B is made to tend to zero.

In conclusion, we see that in the case of concentrated alloys, empirical classifications do not lead to physical models. Needless to say, *the application of such classifications to ceramics or minerals* (e.g. Cannon & Sherby, 1973) *should be avoided.*

4.5   High-temperature deformation of ceramics and minerals

4.5.1   Generalities

Ceramics and minerals are ionocovalent compounds differing only in their origin: ceramics are artificial and minerals natural. Ceramics can be made as pure as possible, whereas there is no such thing as a 'pure' mineral: the ceramic forsterite has the formula Mg_2SiO_4 and Fe ions can substitute for Mg as an impurity, but natural $(Mg, Fe)_2SiO_4$ is the mineral olivine, which can be seen as a solid solution between two end members: Mg_2SiO_4 and Fe_2SiO_4.

The specificity of ceramics and minerals resides in their having *several sublattices, partially or totally occupied by ions of various natures, sizes and electrical charges*. The crystal structure is the result of a compromise between steric and electrostatic requirements; it determines the nature of point defects, the structure of dislocations, as well as the possible interaction between these defects (see Mitchell *et al.*, 1979 for a review).

(i) Point defects: vacancies and interstitials of each species are electrically charged but they can enter neutral associations. There are in general four different types of vacancies which do not necessarily play the same role in high-temperature diffusion:

- Constitutional vacancies are sites which are necessarily empty in a given crystal structure; they do not in general contribute to diffusion since exchanging a constitutional vacancy with an ion creates a defect in the structure.
- Stoichiometry vacancies are vacant sites which involve a departure from the formula unit (e.g. in $Fe_{1-x}O$ there are vacant iron sites and therefore vacant oxygen sites to preserve electrical neutrality); in oxides, the concentration of stoichiometry vacancies depends on the partial pressure of oxygen.
- Extrinsic vacancies are created to preserve electrical neutrality when the formula cations are replaced by aliovalent cations (with different charge but about the same size).
- Intrinsic or thermal vacancies, whose concentration depends on temperature. Some of the consequences of this variety of point defects on diffusion have been reviewed in §2.2.3.

(ii) Dislocations: owing to the fact that the unit cell of ceramics and minerals is often of large dimensions, the Burgers vector of perfect dislocation is large, hence the dislocations are very energetic and tend to dissociate to lower their energy; the existence of several sublattices gives rise to a rich variety of dissociation schemes and stacking faults.

In most oxides, the oxygen ions form a close-packed framework and the glide dislocations dissociate into partials whose motion would cause the cations in the layers between the oxygen close-packed planes to move into sites with the wrong coordination (e.g. tetrahedral instead of octahedral). The cations therefore have to relocate themselves into proper sites and this may be effected by a motion synchronized with that of the dislocation (although not necessarily in the same direction). This process, known as *synchroshear*, was first proposed by Kronberg (1957) for basal slip in Al_2O_3 (fig. 4.17) and later by Hornstra (1960) for $\{111\}$ slip in the spinel structure. Even for unextended dislocations, the core structure is complicated, kinks and jogs are electrically charged (fig. 4.18), which often imparts a net electrical charge to the dislocation. This results in the dislocation being surrounded by a screening atmosphere of oppositely charged point defects. The interaction (elastic or electrical) of point defects with the dislocation core may modify the stoichiometry around the core (or accommodate the non-stoichiometry).

Climb is a complicated process: for instance, the motion by one unit cell vector of a jog in an edge dislocation of olivine Mg_2SiO_4 whose unit cell comprises 4 formula units, involves the transport of 16 oxygen ions, 8 magnesium ions and 4 silicon ions. *Climb is controlled by the diffusion of the slower species, which despite common belief is not necessarily the bulkier anion* (e.g. in olivine Si^{4+} diffuses more slowly than O^{2-}, see table 2.1).

The high-temperature plastic properties of oxides have been reviewed by Bretheau *et al.* (1979) (their tables 4 and 6 give a compilation of creep experiments on oxides). As it would be out of the scope of this book to review here the plastic properties of ceramics and minerals (see Paterson, in preparation), we will only deal briefly here with three specific cases which illustrate the kind of problems met in oxides and silicates: (i) climb dissociation of dislocations and its consequences, (ii) high-temperature creep of olivine single crystals, (iii) creep mechanism in hydrogen oxide ice.

4.5.2   Climb dissociation of dislocations

The crystalline structure of many oxides is such that certain dislocations can easily dissociate in a plane other than the glide plane. This type of dissociation has been particularly investigated in the spinel structure, which we will use as an example.

The spinel structure AB_2O_4 is a very widespread structure that can be analysed as a face-centred cubic (cubic close-packed) framework of oxygen ions whose tetrahedral and octahedral interstices are partly filled with small metallic cations with different electrical charges (Smith, 1982) (fig. 4.19).

Among the most important spinels we can mention: spinel $Mg^{2+}Al_2^{3+}O_4$, magnetite $Fe^{2+}Fe_2^{3+}O_4$, chromite $Fe^{2+}Cr_2^{3+}O_4$,... and the high-pressure silicate phase of the Earth's upper mantle – $Mg_2^{2+}Si^{4+}O_4$. The formula may be written in a different way, which makes the degree of non-stoichiometry apparent: e.g. MgO nAl_2O_3, where $n = 1$ for stoichiometric spinel; n can take values larger than 1 (e.g. 2 or 3). Dislocations with Burgers vector $\mathbf{b} = \frac{1}{2}\langle 110 \rangle$ can dissociate into two partials by the reaction (fig. 4.20):

$$\tfrac{1}{2}\langle 110 \rangle \longrightarrow \tfrac{1}{4}\langle 110 \rangle + \tfrac{1}{4}\langle 110 \rangle$$

Fig. 4.17. Synchroshear in Al_2O_3. (*a*) Dissociation of a dislocation with $\langle 11\bar{2}0 \rangle$ Burgers vector into quarter partials. (*b*) Positions of two layers of oxide ions and an octahedral site in between the layers (A) before and after quarter partial slip. Solid arrows represent the displacement of the upper layer, the dotted arrow indicates the motion of the atom in A (following the octahedral site in its new location) during slip (Kronberg, 1957).

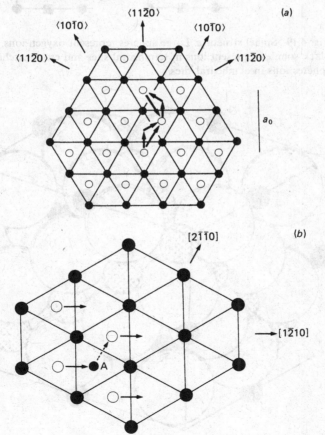

The corresponding stacking fault may lie in the {111} slip plane (glissile fault) but another configuration is possible in which the stacking fault lies in the climb plane (110). The fault in (110) plane with displacement vector $\mathbf{R} = \frac{1}{4}[110]$ has a very low specific energy (Veyssière *et al.*, 1975) as it does not affect the oxygen sublattice (fig. 4.21). The dissociation in the climb

Fig. 4.18. Charged jogs on an edge dislocation in NaCl. Edge dislocation with $\mathbf{b} = \frac{1}{2}[110]$, along [001]. The double extra half plane lies parallel to the plane of the paper (open symbols below the plane). Three possible configurations: negative jog (left), neutral (middle), positive (right) (after Poirier, 1972). Circles – cations; squares – anions.

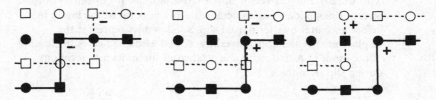

Fig. 4.19. Spinel structure. Large spheres represent oxygen ions, small black spheres represent ions in tetrahedral sites and cross-hatched spheres ions in octahedral sites.

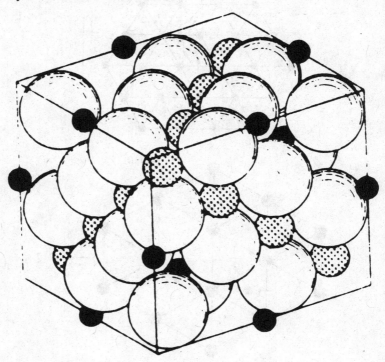

Fig. 4.20. Dissociation of a dislocation with $\frac{1}{2}\langle 110\rangle$ Burgers vector in the spinel structure (after Bretheau *et al.*, 1979).

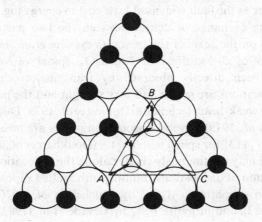

Fig. 4.21. Stacking fault $\frac{1}{4}[110]$ (110) in the spinel structure. The oxygen ions are not represented and the metallic ions are represented inside their coordination polyhedra (tetrahedra and octahedra) (Madon & Poirier, 1980).

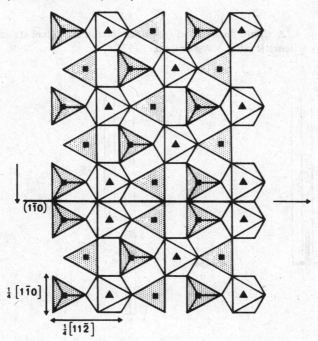

plane can take place by climb of the two partials in opposite directions, one of the partials absorbing the vacancies created at the others. The partials recede from each other as the fault widens at little cost in energy (fig. 4.22). The process involving exchange of matter between the two partials is globally conservative on the scale of the crystal (Veyssière *et al.*, 1978).

Climb dissociation of dislocations in $MgAl_2O_4$ spinel of various stoichiometries has been directly observed by transmission electron microscopy: the dislocations are sessile and very straight and the partials can be resolved by the weak-beam technique (Donlon *et al.*, 1979; Duclos *et al.*, 1978; Doukhan *et al.*, 1979). Several dissociation planes are possible – $\{100\}$, $\{110\}$ and even $\{113\}$ for spinel with $n = 1.8$ (Doukhan *et al.*, 1982); the dissociation plane may continuously change along the dislocation (fig. 4.23). Climb dissociation of dislocations resulting into faulted dipoles has also been observed in sapphire (α-Al_2O_3) by Mitchell *et al.* (1976) and Phillips & Cadoz (1982), and in pyrite FeS_2 by Levade *et al.* (1982) (figs. 4.24, 4.25). Climb dissociation of the dislocations makes them sessile and a dislocation that has reached an orientation in which it can split in its climb plane should in principle be prevented from further slipping. Under these conditions, creep strain may, however, result from the operation of the following processes:

Fig. 4.22. Climb dissociation of dislocations in the spinel structure. (*a*) Schematic (after Veyssière *et al.*, 1978).

(*a*)

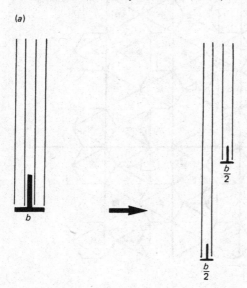

(i) Pure climb of the dislocations in their climb planes, as proposed by Duclos *et al.* (1978) for almost stoichiometric spinel. This process is also thought to be rate-controlling in the high-temperature creep (1550 < *T* < 1800 °C) of yttrium oxide (Y_2O_3 (Gaboriaud, 1981); this oxide has a cubic

(*b*) Configuration (Doukhan *et al.*, 1982).

(*b*)

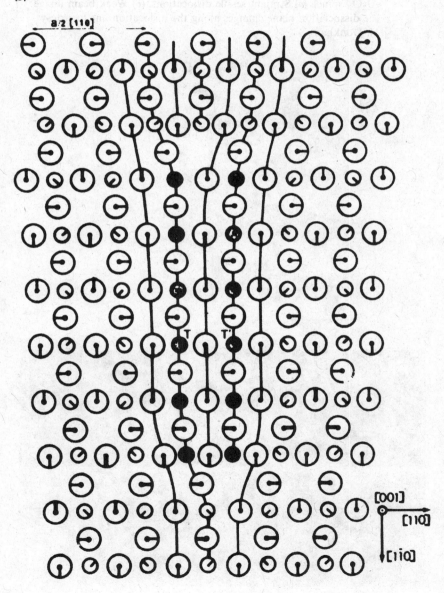

structure, related to that of fluorite CaF_2 and the dislocations with $\langle 100 \rangle$ Burgers vector may split in their climb planes with a fault of low energy; climb is controlled by the diffusion of the slower ion, in this case the Y^{3+} cation.

Fig. 4.23. TEM of dislocations dissociated in climb in MgO 1.8% Al_2O_3 spinel. (*a*) Straight sessile dislocations. (*b*) Weak beam image: the dissociation plane changes along the dislocation line (courtesy N. Doukhan).

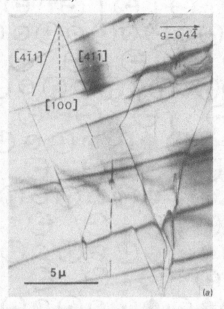

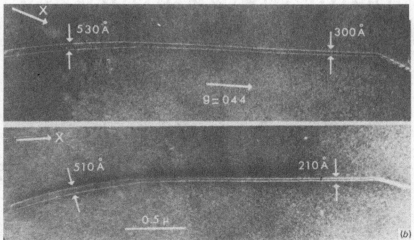

(ii) Viscous glide of the split dislocations in their glide planes, controlled by the 'diffusion' of the fault ribbon (Duclos *et al.*, 1982). Stoichiometry vacancies in spinels with high *n* facilitate the cationic rearrangements required by the motion of the fault normal to its plane.

As climb dissociation of dislocations is controlled by diffusion, it must become easier as temperature increases; this may lead to an unusual variation of the mechanical properties with temperature (fig. 4.26): In constant-strain-rate tests, the flow stress normally decreases as temperature

Fig. 4.24. TEM of dipole loop in Al_2O_3 deformed at 1450 °C (scale bar = 0.1 µm) (*c* axis along the arrow). The dissociation plane varies along the loop (courtesy J. Rivière & J. Cadoz).

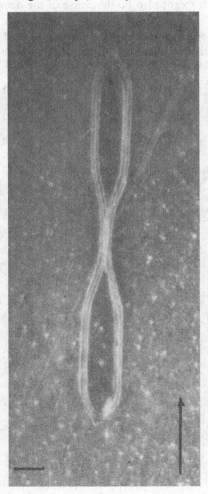

increases up to a certain temperature, then the flow stress starts increasing with temperature until its variation becomes 'normal' again and it starts decreasing. This behaviour was first recognized in pure magnesium single crystals deformed by slip on the pyramidal $\{11\bar{2}2\}\langle11\bar{2}3\rangle$ system (Stohr & Poirier, 1972); it was explained by the thermally activated splitting of the $c + a$ edge dislocations out of their glide plane (fig. 4.27). Climb splitting of c dislocations in the basal plane may well be the cause of the total absence of c [0001] glide in all hexagonal metals while pure climb of c dislocation may cause considerable strain (Edelin & Poirier, 1973). A remarkable increase of the flow stress with temperature has recently been observed in andalusite Al_2SiO_5 between 500 °C and 900 °C (fig. 4.28) (Doukhan & Paquet, 1982); it was attributed to the dissociation of the [001] dislocations out of the (110) glide plane. Climb dissociation might also be the cause of the decrease in

Fig. 4.25. TEM of dislocations dissociated in climb in pyrite FeS_2. Dislocation with $\mathbf{b} = [010]$. (*a*) The edge portion arrows are dissociated in climb in (010); (*b*) the mixed portion arrows are dissociated in glide. (Courtesy C. Levade.)

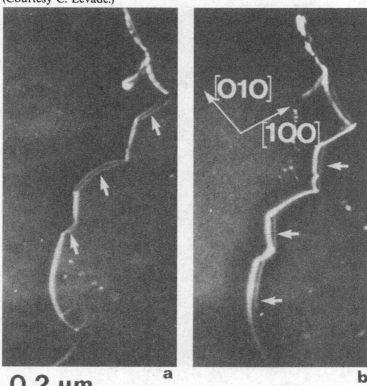

0.2 μm

creep-rate as temperature increases observed in fluoperovskite $KZnF_3$ between 750 °C and 800 °C (Poirier *et al.*, 1983).

It is quite probable that, as investigations on ionocovalent crystals become more numerous, the 'unusual' temperature dependence of flow

Fig. 4.26. Hardening with temperature caused by climb dissociation. Solid curve: stress at constant strain-rate. Dashed curve: creep-rate at constant stress. A: Normal behaviour with increasing *T*. B: Slip becomes more difficult due to thermally activated climb dissociation. C: Temperature becomes high enough for stacking faults to diffuse or for some other process to take over.

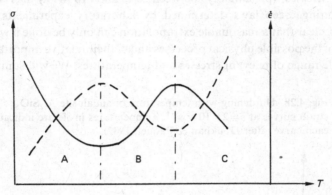

Fig. 4.27. Climb dissociation of c + a edge dislocations in magnesium (TEM) (scale bar = 0.5 μm) (Stohr & Poirier, 1972).

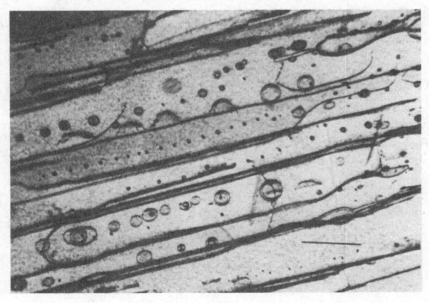

stress or creep-rate will become more and more usual and that climb dissociation of dislocations will be found responsible in many cases.

4.5.3   Creep of olivine single crystals

Olivine $(Mg, Fe)_2SiO_4$ is an essential constituent of the Earth's upper mantle, down to a depth of about 400 km. High-temperature creep processes therefore control the convection in the mantle, which explains the interest geophysicists take in the investigations of creep in olivine. The need for a rheological law valid at very low strain-rates ($\dot{\varepsilon} \approx 10^{-14}/s$) and low applied stresses (presumably between 0.1 and 10 MPa) has led to extrapolating creep laws determined by laboratory experiments. It is, however, clear that a reasonable extrapolation can only be done if we have an idea of the possible physical processes and of their relative importance in the whole range of relevant stresses and temperatures. We will summarily

Fig. 4.28. Hardening with temperature of andalusite Al_2SiO_5: Stress–strain curves at $\dot{\varepsilon} = 2 \times 10^{-5}$ s^{-1}. Temperatures in °C are indicated on each curve (after Doukhan & Paquet, 1982).

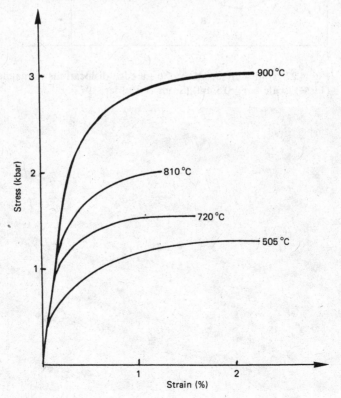

review the evidence which points to the presence of both climb- and glide-controlled creep, but is not yet sufficient for definite conclusions to be drawn. Olivine is a generic name for a series of minerals with continuous substitution of Fe for Mg, from the magnesian end member, forsterite (Mg_2SiO_4), to the ferrous end member, fayalite (Fe_2SiO_4). The oxygen ions form a quasi-hexagonal-close-packed framework, but the cation filling gives the structure an orthorhombic symmetry. The point defect chemistry has been reviewed by Smyth & Stocker (1975) and the slip systems have been reviewed and analysed in terms of the HCP lattice of oxygens by Poirier (1975) (fig. 4.29).

Numerous experiments have been conducted on natural olivine rocks (dunites) under pressure (see Nicolas & Poirier, 1976, and Paterson, in preparation) but as we are interested here in the physical process, we will only review the creep experiments on single crystals at ambient pressure. All experiments were performed in dead-weight-loading uniaxial creep on samples of various orientations; the results were analysed in terms of

Fig. 4.29. HCP lattice of oxygen ions in olivine. Crystallographic directions are given in Miller indices (orthorhombic system) and Miller–Bravais indices (four indices) for HCP system. The four SiO_4 tetrahedra of a unit cell are represented (after Poirier, 1975).

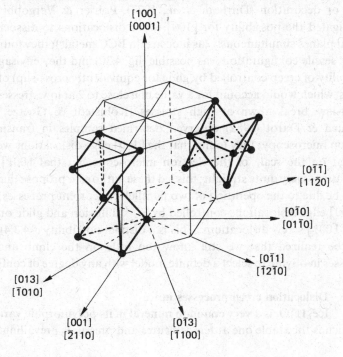

power-law creep. The experiments were conducted on natural San Carlos olivine with 92 mol % forsterite (Kohlstedt & Goetze, 1974; Durham *et al.*, 1977) and on synthetic forsterite, which has the advantage that creep is insensitive to the oxygen partial pressure (Durham *et al.*, 1979; Darot & Gueguen, 1981). The results are summarized in table 4.2. It appears, once more, that the mere knowledge of values of n and Q is completely insufficient to narrow down the field of possible mechanisms. Analysis of the shape change of the deformed specimens led Durham & Goetze (1977a) to ascribe a part of the creep strain to climb of dislocations; however, the activation energy for creep is always larger than the activation energy for diffusion of oxygen or silicon (≈ 90 kcal/mol) (Jaoul *et al.*, 1981; Jaoul & Houlier, 1983).

The observation of dislocation structures in deformed specimens by a decoration technique and by optical microscopy (Kohlstedt *et al.*, 1976) revealed various characteristic configurations according to the orientation of the samples and the active slip systems, in particular apparently straight edge dislocations in tilt boundaries when the (010) [100] slip system is active (Durham *et al.*, 1977). Despite its interest, the decoration technique does not allow the determination of the dislocation Burgers vectors and sometimes led to gratuitous assumptions. On the basis of observations of straight [100] screw dislocations by electron microscopy (Phakey *et al.*, 1972) or decoration (Durham *et al.*, 1977), Poirier & Vergobbi (1978) investigated the possibility for [100] screw dislocations to dissociate on several planes simultaneously (as it occurs in BCC metals); they found that such a sessile configuration was possible (fig. 4.30) and they envisaged the possibility of creep controlled by glide (or equivalently cross-slip) of [100] screws, which would account for a value of n close to 2 at low stresses and a power-law break-down at high stresses (Kohlstedt & Goetze, 1974). Gueguen & Darot (1982) observed deformed samples in transmission electron microscopy; they found that the [100] edge dislocations were not straight on the scale of the electron microscope but that [001] screw dislocations were quite straight; this led these authors to propose that creep might be due to the operation of two parallel–concurrent processes: climb of [100] edge dislocations controlled by Si^{4+} diffusion and glide or cross-slip of [001] screw dislocations. This is a distinct possibility (§ 4.3.4), but it must be realized that we still know too little on the climb and glide processes in olivine to select a definite model with any degree of confidence.

4.5.4   Dislocation creep processes in ice

Ice, H_2O, is a very common mineral in its polymorphic variety ice I_h, which is the stable one at temperatures and pressures prevailing on the

Table 4.2. Creep experiments on single crystals of olivine

T (°C)	σ (MPa)	Orientation	Composition	n	Q (kcal/mol)	Reference
1430–1650	5–150	Various	Fo 92	varies with σ	125	Kohlstedt & Goetze, 1974
1400–1600	10–60	B	Fo 92	3.7	125	Durham & Goetze, 1977a
1500–1600	10–40	D	Fo 92	3.5	125	Durham & Goetze, 1977a
1550–1650	15–50	B	Fo 100	2.6	140	Durham & Goetze, 1977b
1480–1680	3–30	D	Fo 100	2.9	160	Durham et al, 1979
1400–1650	10–100	A	Fo 100	2.6	110	Darot & Gueguen, 1981
1400–1600	20–100	B	Fo 100	3.6	137	Darot & Gueguen, 1981
1500–1610	30–100	C	Fo 100	2.7	143	Darot & Gueguen, 1981

Orientations: A, B, C – the stress axis makes an angle of 45° with [100] and [010] and [001], respectively; D – the stress axis makes equal angles with [100], [010] and [001].

Earth's surface ($-50\,^{\circ}$C $< T < 0\,^{\circ}$C; $1 < P < 200$ bar). Other polymorphic varieties at low temperatures and/or higher pressures exist as primary constituents of the ice moons of the outer planets Jupiter and Saturn (see Poirier, 1982). The peculiar properties of hydrogen bonds give rise to special processes controlling dislocation motion in ice.

(i) Hexagonal ice I_h has a würtzite-type structure: the oxygen ions form layers of puckered hexagonal rings stacked in an ABAB sequence (fig. 4.31); the O^{2-} ions are tetrahedrally coordinated. Protons forming the bonds between the oxygens are disordered but follow Bernal–Fowler rules: there must be two protons close to any oxygen ion and only one proton per bond. Infringement of these rules results in point defects specific to the ice structure: the Bjerrum defects, consisting in bonds with no proton (L defect) or two protons (D defect). A dislocation moving through the lattice necessarily creates Bjerrum defects in the array of disordered protons obeying the Bernal–Fowler rules (Glen, 1968) (fig. 4.32); as the energy of these defects is quite high (0.64 eV/atom) this would result in a very large Peierls stress close to the theoretical elastic limit ($\approx 0.1\mu$) unless the dislocation meets diffusing Bjerrum defects, which rearrange the protons ahead of the dislocation and allow the kinks to move sideways without having to create the defects. The dislocations are therefore subjected to a

Fig. 4.30. Core of a [100] sessile dislocation in olivine, split on (010) and (001) planes (hard sphere model). The ions are in unrelaxed positions (after Poirier & Vergobbi, 1978).

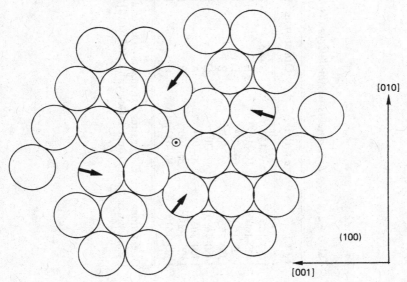

[010]

(100)

[001]

Fig. 4.31. Ice I_h structure. Puckered hexagonal rings of oxygen ions stacked in an ABAB sequence. There are two sets of basal planes: the shuffle set (*S*) and the glide set (*G*) (after Whitworth, 1978).

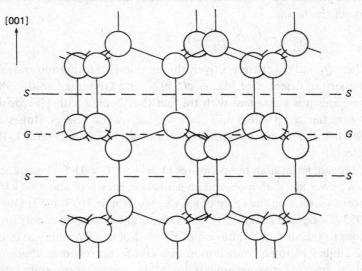

Fig. 4.32. Motions of a dislocation in ice creates Bjerrum defects. (*a*) Perfect lattice obeying the Bernal–Fowler rules. (The trace of the basal planes is horizontal.) (*b*) Edge dislocation moving on the shuffle set *S*. (*c*) After the passage of the dislocation, there is a D-defect on BC′ bond (2 protons) and an L-defect on CD′ bond (0 proton).

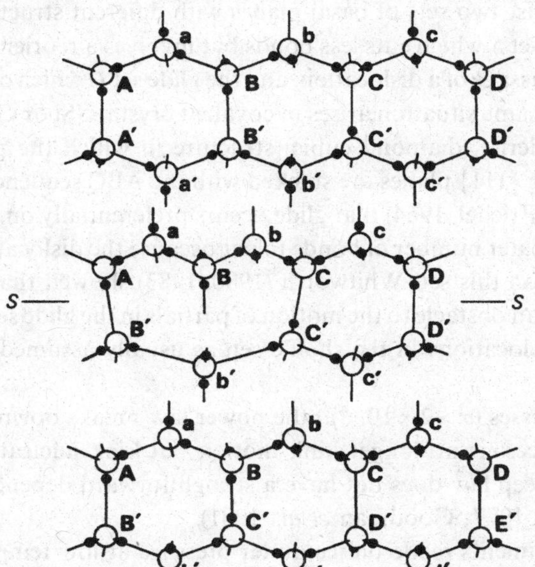

viscous drag and their motion is said to be controlled by proton rearrangement; this process occurs at low stresses. With the usual assumption of a σ^2 dependence of the dislocation density, we obtain a creep law of the form:

$$\dot{\varepsilon} \propto \sigma^3 \exp\left(-\frac{Q_F + Q_M}{kT}\right) \qquad (4.45)$$

where Q_F and Q_M are respectively the enthalpies of formation and migration of Bjerrum defects (see Maï *et al.*, 1977, and Goodman *et al.*, 1981). This mechanism is consistent with the fact that doping with HF considerably lowers the creep stress and the creep activation energy (Jones & Glen, 1969): it is known that the dissolution of HF in ice creates L-Bjerrum-defects.

Indeed, ice at high temperatures ($T > -50\ ^\circ C \approx 0.8 T_m$) creeps according to a power law with $n = 3$ and an activation energy of about 60 kJ/mol (for recent reviews on the creep of ice, see Weertman, 1973, 1983; Duval *et al.*, 1983; Goodman *et al.*, 1981). The value of the activation energy for creep is close to the value for diffusion of H^+ and of oxygen; this leaves open the possibility of some contribution of a climb-controlled mechanism to the creep rate, but the arguments for glide control seem quite compelling (Weertman, 1983). It seems, however, that the dislocation velocities calculated for the proton-rearrangement mechanism are much lower than the velocities observed experimentally for basal glide (the easy slip system in ice). This discrepancy has been attributed by Whitworth (1980, 1983) to the fact that there exist two sets of basal planes with different structures (fig. 4.31): the shuffle set S which cuts less bonds but involves a reorientation of bonds after the passage of a dislocation, and the glide set G which cuts more bonds. Now, the same situation arises in covalent crystals (Si or Ge) which possess the sphalerite (diamond cubic) structure in which the puckered hexagons forming $\{111\}$ planes are stacked with the ABC sequence, and it has been shown (Friedel, 1964) that glide occurs preferentially on the glide set despite the greater number of bonds to cut because the dislocations can easily dissociate on this set. Whitworth (1980, 1983) showed that proton disorder is less of an obstacle to the motion of partials in the glide set than to the motion of dislocations in the shuffle set, as usually assumed (e.g. fig. 4.32).

For higher stresses ($\sigma > 2 \times 10^{-4} \mu$) the power law breaks down and the rate-limiting process is no longer kink motion but kink nucleation. The corresponding creep law does not have a straightforward dependence on stress (Maï *et al.*, 1977; Goodman *et al.*, 1981).

The first experiments made on ice under pressure at low temperatures (Durham *et al.*, 1984) show that, even though the stress exponent is $n = 4$

over the whole temperature range investigated (158 K < T < 268 K), the apparent activation energies for creep take different values in three domains:

31 kJ/mol for 158 K < T < 195 K
61 kJ/mol for 195 K < T < 243 K
91 kJ/mol for 243 K < T < 268 K

The result is that ice deforms more easily at low temperatures than would be predicted by extrapolation of results obtained above − 50 °C (fig. 4.33).

(ii) The creep properties of the other polymorphic varieties of ice are almost totally uninvestigated, except for ice VI whose viscosity has been measured in a sapphire-anvil cell (Poirier *et al.*, 1981). The interesting crystalline structures of the high-pressure varieties of ice (see Parsonage & Staveley, 1978) permit some speculation on the possible dislocation processes. In particular, the densest form of ice, ice VII, stable above about 20 kbar and 0 °C, has an oxygen lattice that can be described as two interpenetrating sphalerite (or diamond cubic) lattices with no hydrogen bonds between them. In this case, any dislocation gliding on a {111} plane must use the shuffle set of at least one lattice (fig. 4.34); as the protons are

Fig. 4.33. Arrhenius plot for the flow stress of ice I_h ($P = 50$ MPa, $\dot{\varepsilon} = 3.5 \times 10^{-6}$/s). The activation energy falls from 91 kJ/mol near the melting point to 31 kJ/mol at − 115 °C (after Durham *et al.*, 1984). From top to bottom $\dot{\varepsilon} = 3.5 \times 10^{-4}$, 3.5×10^{-5}, 3.5×10^{-6}/s.

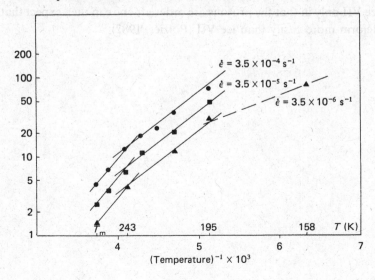

Fig. 4.34. Structure of ice VII: Two interpenetrating diamond cubic lattices (projection on (110) plane). Glide on (111) planes necessarily involves the shuffle set of one of the lattices: 1 and 3 – shuffle sets of lattices A and B; 2 – shuffle set of lattice A, glide set of lattice B; 4 – glide set of lattice A, shuffle set of lattice B.

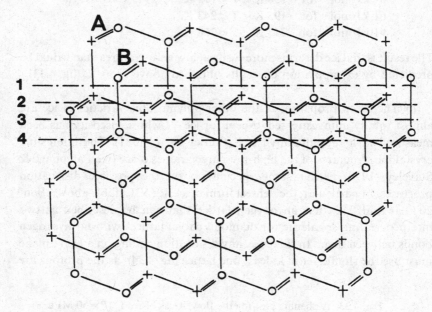

disordered, one can expect a high Peierls stress for glide in ice VII. The structure of ice VIII (below 0 °C, but still above 20 kbar) differs from that of ice VII only in that the protons are ordered; one can then expect that it will deform more easily than ice VII (Poirier, 1982).

5

The effect of hydrostatic pressure on deformation

The effect of hydrostatic pressure on deformation is of secondary interest in materials science but it is very important in geophysics, as the hot silicate Earth's mantle extends down to depths where the hydrostatic pressure reaches values higher than 1 Mbar. The structure and composition of the Earth are summarily presented according to the latest models, and it is shown that the possibility of creep, hence convection in the lower mantle, is directly linked to the variation of the activation volume with pressure.

The effect of pressure on mechanical properties, either diffusion- or glide-controlled, is shown, in most cases, to be a manifestation of the anharmonicity of the crystals. It can be expressed in terms of the variation of the elastic constants with pressure. The effect is small in the case of athermal processes but may be large for the thermally activated processes, for which the energy barrier can be elastically modelled. Another indirect effect has its source in the dependence on pressure of the diffusivity of impurities (such as water-related species in quartz).

Elastic models for the activation volume of creep are reviewed and applied to the Earth's lower mantle.

5.1   Generalities: high-pressure deformation and geodynamics

The effect of hydrostatic pressure P on the flow stress or creep-rate of non-porous materials is much less important than the effect of temperature: indeed, for values of P currently obtained in the laboratory (a few tens of kilobars), there is only a second-order difference between the creep-rate under pressure and at one atmosphere pressure. As a consequence, after a few exploratory experiments, the materials science community has, by and large, lost interest in the field of deformation under pressure: the scientific or technological return was too small for a high experimental investment. In geophysics, however, the rheological properties of minerals and rocks at high temperatures and pressures inside the Earth constitute a topic of prime importance; as pressure increases more rapidly than temperature with depth and reaches very high values (fig.

5.1), its role in flow processes cannot be neglected. After a brief and sketchy survey of the Earth's structure and composition and of the relevant rheological problems, we will review the physical basis of the effect of pressure on flow stress and creep-rate.

5.1.1   The structure and composition of the Earth

Despite some obviously exaggerated claims (Verne, 1864), there has been no exploration of the inside of the Earth, and we must therefore rely on Earth models to obtain some information as to its structure and composition.

Seismological models use the propagation of seismic (elastic) waves through the Earth to obtain profiles of sound velocity, density and elastic constants with depth (table 5.1). The Earth can be divided into three concentric shells where the sound velocity varies more or less smoothly

Fig. 5.1. Variation of pressure and possible variation of temperature with depth in the Earth. 1. $T(z)$ from Anderson (1982). 2. $T(z)$ from Jeanloz & Richter (1979). 3. $P(z)$ from Anderson (1982) (dashed curve).

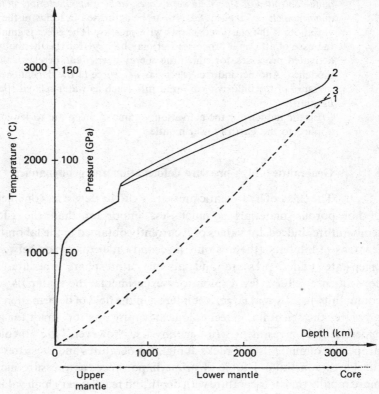

with depth, separated by sharp discontinuities: the crust, the mantle and the core. Elastic equations of state are used to extrapolate back the densities and moduli to atmospheric pressure; these quantities can then be compared with the values for mixtures of known minerals or of minerals stable only at very high pressures and obtained by laboratory synthesis. The outcome is a petrological Earth model. Current Earth models differ only in details and for our present purposes it is sufficient to describe the Earth as follows (table 5.1):

(i) The *crust* extends to a depth of about 40 km under the continents; it is therefore only a thin veneer on the Earth, whose mean radius is 6371 km. The continental crust is light and has a mostly granitic composition: the

Table 5.1. *Seismological elastic Earth model*

Depth z (km)	Density ρ	Pressure P (Mbar)	Bulk modulus B (Mbar)	Shear modulus μ (Mbar)
Crust				
10	2.8	0.0003	0.54	0.35
Upper mantle				
50	3.3	0.015	1.13	0.72
100	3.35	0.025	1.25	0.63
200	3.4	0.072	1.28	0.65
420	3.55	0.141	1.76	0.82
420	3.77	0.141	2.16	0.96
570	3.95	0.199	2.35	1.11
670	4.08	0.239	2.48	1.22
Lower mantle				
670	4.38	0.239	3.05	1.64
1000	4.57	0.387	3.54	1.85
1500	4.85	0.621	4.32	2.15
2000	5.12	0.869	5.13	2.44
2500	5.37	1.135	5.95	2.71
2890	5.55	1.354	6.35	2.91
Outer core				
2890	9.91	1.354	6.58	0
4000	11.32	2.461	10.31	0
5150	12.14	3.289	12.77	0
Inner core				
5150	12.71	3.289	13.63	1.50
6370	13.01	3.632	14.24	1.65

After Dziewonski *et al.*, 1975.

main minerals are SiO_2 as quartz and alkali feldspars $(K, Na)AlSi_3O_8$. The pressure in the lower crust is of the order of 10 kbar and the temperature about 700 to 800 °C. The rocks of the lower crust undergo plastic deformation during mountain building and their rheological properties are actively investigated in laboratory experiments where pressure is only a means of preventing early fracture and achieving plastic strain (see Paterson, 1978 and in preparation).

(ii) The *mantle* extends from the base of the crust down to 2900 km, it therefore constitutes about 80% of the total volume of the Earth. In first approximation, its chemical composition can be thought constant throughout its thickness; it is essentially that of a peridotite composed in equal molar proportions (Liu, 1979) of olivine

$$[Mg_2SiO_4]_{0.9}[Fe_2SiO_4]_{0.1}$$

and aluminous pyroxene

$$[MgSiO_3]_{0.81}[FeSiO_3]_{0.09}[Al_2O_3]_{0.1}$$

The mantle can be divided into two regions, separated by a seismic discontinuity, and corresponding to the stability domains of different mineral structures with the same chemical composition (Liu, 1979; Bassett, 1979; Bell, 1979).

The *upper mantle* extends down to about 700 km; its principal minerals are olivine $(Mg, Fe)_2SiO_4$ and pyroxene $(Mg, Fe)SiO_3$, stable at atmospheric pressure and which transform into denser phases as the pressure increases in the transition zone (between 400 and 700 km). Olivine transforms into spinel $(Mg, Fe)_2SiO_4$ and pyroxene transforms into garnet; the silicon ions remain 4-coordinated. The 700-km seismic discontinuity between upper and *lower mantle* corresponds to phase transitions to the densest silicate structures synthesized so far: the $(Mg, Fe)SiO_3$ minerals take the perovskite structure (fig. 5.2) where silicon is 6-coordinated $(SiO_6$ octahedra instead of SiO_4 tetrahedra at lower pressures); the spinel transforms into perovskite and magnesiowüstite:

$$(Mg, Fe)_2SiO_4 \rightarrow (Mg, Fe)SiO_3 + (Mg, Fe)O$$

As a result, perovskite $(Mg, Fe)SiO_3$ is the principal constituent of the lower mantle.

(iii) The *core* is essentially an iron–nickel alloy; it is mostly in the liquid state, except near the centre, and it will not concern us here.

5.1.2   Convection in the viscous mantle

The hot, crystalline silicate mantle is the seat of convection currents, i.e. it undergoes high-temperature creep. That this is so results from the constraints of plate tectonics on an Earth that does not expand (there must be inner return currents to close the cells that drive the plates apart) and from the constraints of plausible thermal models (the internal heat cannot be evacuated by conduction alone, convection must take place) (see Uyeda, 1978). The necessity of high-temperature creep of the silicate mantle was rapidly recognized, and rheological laws for mantle materials were extrapolated from laboratory experiments on olivine and peridotites (McKenzie, 1967; Gordon, 1971; Stocker & Ashby, 1973*b*; Weertman & Weertman, 1975; Weertman, 1978; Ashby & Verrall, 1978; Gueguen & Nicolas, 1980). Profiles of the variation of viscosity with depth (fig. 5.3) were constructed for the upper mantle, usually on the basis of diffusion-controlled creep mechanisms. Thus, it was predicted that viscosity would first decrease and then increase with depth; this, of course, is what can be expected from any diffusion-controlled creep law with given values of the activation energy and of the activation volume, when the temperature and pressure vary with depth as in fig. 5.1: the temperature increases faster than pressure at first, causing an overall increase in the creep-rate, but below a depth of about 100 km the temperature increases slowly whereas pressure keeps on increasing at roughly the same rate and reaches high values, eventually more than cancelling the softening effect of temperature (fig. 5.3). These results were extrapolated to the lower mantle, assuming a similar creep law for its material and an activation volume independent of

Fig. 5.2. Perovskite structure ($MgSiO_3$). Mg^{2+} ions: empty circles. O^{2-} ions: stippled circles. Si^{4+} ion: solid circle, in the octahedron of oxygen ions. Mg and O ions are the same size and form together an FCC lattice.

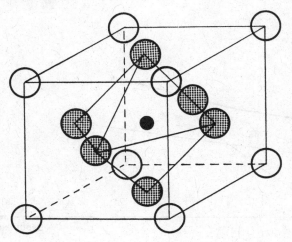

pressure; the lower-mantle viscosity was then usually thought to increase to very high values, precluding convection, despite the fact that the analysis of the viscoelastic postglacial rebound of the Scandinavian and Canadian shields was compatible with an almost constant viscosity of 10^{21}–10^{22} poise throughout the whole mantle. The view that convection takes place only in the upper mantle has recently come under attack, and the lower mantle is now seen by many geophysicists as convecting, either together with the upper mantle or separately (O'Connell, 1977; Davies, 1977; Elsasser *et al.*, 1979; Jeanloz & Richter, 1979; Richter & McKenzie, 1981; Spohn & Schubert, 1982; Peltier & Jarvis, 1982). Clearly, the problem of finding the variation of viscosity of the mantle with depth is one involving the pressure dependence of creep-rate and we will examine it in this light in § 5.3.1, but it must be kept in mind that the variation of viscosity with depth also depends strongly on the temperature profile $T(z)$ (geotherm), which in turn depends on whether or not the viscosity is low enough to allow the heat to be evacuated by convection.

Fig. 5.3. Qualitative variation of viscosity with depth in the Earth expected for a diffusion-controlled mantle creep law: $\eta \propto$ $\exp\left[(Q+P\Delta V)/RT\right]$ and constant activation volume ΔV. The profiles $P(z)$ and $T(z)$ are qualitatively reasonable.

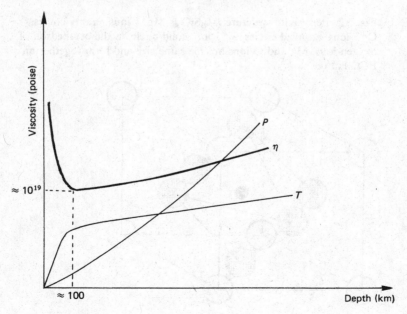

5.2 The physical basis of the effect of pressure on flow stress

As hydrostatic pressure does not cause shear strain, its effect on flow can only be indirect: it must originate in a volume variation ΔV at some stage in the chain of processes leading to the flow stress or creep-rate.

If we refer to Orowan's equation $\dot{\varepsilon} = \rho v b$, we see that pressure can affect it in two ways: through the dislocation density or through the dislocation velocity (in glide or climb). In many cases, the effect of pressure can be simply ascribed to anharmonicity of the crystals, i.e. to the variation of the elastic constants with pressure.

5.2.1 Effect of pressure on dislocation creation

Pressure can affect dislocation creation and multiplication in several ways.

(i) In non-cubic crystals, the linear compressibility is anisotropic, i.e. a single crystal subjected to hydrostatic pressure will not contract by an equal amount in all directions. The application of hydrostatic pressure P to a polycrystal of such a material will cause internal stresses at grain boundaries. Paterson (1963) calculated that for calcite the local normal stresses on grains vary from $0.82P$ parallel to c-axis to $1.09P$ perpendicular to c-axis. If the temperature is high enough, the local stresses may relax by creation of dislocations and plastic deformation near the grain boundaries, even though no shear stress is applied externally (Gelles, 1966). This effect is obviously transient and occurs only during the establishment or changes of pressure, its influence on the dislocation density, hence on the flow stress or creep-rate, of polycrystals deformed under pressure is negligible (Oguchi *et al.*, 1972a). It is, however, possible in some cases to reduce the yield stress by maintaining elastically heterogeneous materials to high pressure prior to deformation (*pressurization*). This was found to be beneficial in the case of brittle chromium which could be deformed to 60% strain after pressurization at 10 kbar (Bullen *et al.*, 1964): mobile dislocations were created at elastic heterogeneities.

(ii) *The introduction of dislocations in a crystal is accompanied by a slight volume expansion* (i.e. a crystal containing dislocations has a lower density than a perfect crystal). This is, in fact, a consequence of anharmonicity: the presence of dislocations causes internal strains in the crystal, with a corresponding stored elastic energy proportional to the elastic moduli; the free energy of the self-strained crystal containing dislocations is therefore lowered by a volume expansion (causing an anharmonic decrease of the elastic moduli). Seeger & Haasen (1958) have calculated the volume expansion δV per unit length of dislocation:

$$\delta V = \frac{1}{B}\left[\frac{dB}{dP}-1\right]E_d + \frac{1}{\mu}\left[\frac{d\mu}{dP}-\frac{\mu}{B}\right]E_s \qquad (5.1)$$

where B and μ are the bulk and shear modulus respectively; E_d and E_s are the dilatation and shear part, respectively, of the dislocation line energy per unit length; they are proportional to $\ln(R/b_0)$ (see eq. (2.56)) where b_0 is the core radius and R the radius of the cylinder around the dislocation where the energy is stored.

For an edge dislocation:

$$E_d = \frac{1}{3}\frac{\mu b^2}{4\pi}\left[\frac{54\mu B}{(3B+4\mu)^2}\right]\ln\left(\frac{R}{b_0}\right)$$

and

$$E_s = \frac{2}{3}\frac{\mu b^2}{4\pi}\left[\frac{9B^2+42\mu b+4\mu^2}{(3B+4\mu)^2}\right]\ln\left(\frac{R}{b_0}\right)$$

If the dislocations are assimilated to parallel lines, with density ρ we have:

$$R = \frac{1}{(\pi\rho)^{1/2}}$$

Taking $b_0 \approx b$, and assuming $\rho \approx 10^8\ \text{cm}^{-2}$, we can calculate the volume expansion ΔV per length b of edge dislocation (the formation volume of a dislocation, so to speak):

for aluminium	$\Delta V = 1.7$ atomic volume
sodium chloride	$\Delta V = 3.9$ molecular volume
forsterite	$\Delta V = 0.9$ molecular volume

This expansion is rather small: in the case of forsterite it corresponds to an overall dilatation of about 10^{-7}. Pressure works against the volume expansion and it must follow that dislocation multiplication is somewhat more difficult under pressure. This must be reflected by an increase of the work-hardening coefficient, hence a slight increase in the flow stress under pressure. Haasen & Lawson (1958) found $\Delta\sigma/\sigma \approx 1\%$ for copper and aluminium under 5 kbar pressure and explained it in this way.

(iii) The critical stress for the activation of Frank–Read sources increases linearly with the shear modulus (eq. (2.58)). If the flow stress is controlled by the production of dislocations at Frank–Read sources it should increase with pressure as the shear modulus.

5.2.2   Effect of pressure on dislocation velocity

(i) *Athermal flow processes*

The stress field of dislocations varies as $\mu b/r$, where μ is the shear modulus and r the distance from the dislocation. Therefore, the stress needed to overcome long-range athermal obstacles, such as the internal stress-field undulations, will linearly depend on the shear modulus: the same is true of the stress needed to overcome obstacles formed by dislocation configurations (junctions, tangles, etc.). Whenever the flow stress is controlled by such a mechanism, we can expect that it will increase with pressure as the shear modulus: the flow stress σ_P at pressure P is related to σ_0 at atmospheric pressure by:

$$\sigma_P = \frac{\mu_P}{\mu_0}\,\sigma_0 = \frac{\mu_0 + \dfrac{\mathrm{d}\mu}{\mathrm{d}P}\,P}{\mu_0}\,\sigma_0 = \left(1 + \frac{\mathrm{d}\ln\mu}{\mathrm{d}P}\right)\sigma_0 \qquad (5.2)$$

The effect is important if μ_0 is small: in the case of the metal potassium deformed at constant strain rate at 77 K ($0.2T_\mathrm{m}$) the flow stress increases by 42% for a pressure of 5 kbar (Chua & Ruoff, 1975) (fig. 5.4), in good agreement with an increase in shear modulus of 35%. Other metals deformed at room temperature under 12 kbar exhibit an increase in flow stress comparable to the increase in shear modulus: about 9% for aluminium (Yoshida & Oguchi, 1970), 8% and 3% respectively for zinc and zirconium (Oguchi *et al.*, 1972a) and 2% for pure iron (Oguchi *et al.*, 1972b).

Alkali halide single crystals (KCl, NaCl, KBr, KI, RbI, CsBr, LiF) were tested under a pressure of 4 kbar at room temperature (Davis & Gordon, 1968); whereas crystals hardened by irradiation, in which flow was presumably controlled by elastic interactions, did exhibit an increase in flow stress comparable with the modulus increase, the soft crystals behaved erratically; direct measurement of dislocation mobility under pressure by etch pits in KCl (Haworth & Gordon, 1970) showed no detectable effect of pressure on velocity, in apparent contradiction with the important effect reported on the flow stress (Davis & Gordon, 1968). Single crystals of MgO compressed along [100] at 10 kbar showed a negligible effect of pressure on the flow stress (Auten *et al.*, 1976). On the whole, the experiments performed so far do not lead to any firm conclusion, probably because the crystals were not extremely pure and the flow mechanisms were not clearly identified.

(ii) *Flow controlled by overcoming of Peierls barriers*

Pure nickel wafers deformed at room temperature between rotating anvils under pressures up to 150 kbar exhibited a strong increase in flow stress with pressure, an order of magnitude greater than the modulus increase (Jesser & Kuhlmann-Wilsdorf, 1972). This effect was attributed to a pressure dependence of the Peierls stress: the core of the dislocations expanding and contracting as the dislocation moves from one valley to another ('breathing' of the core). The observed stress increase for 150 kbar corresponds to a dilatation of 10^{-2} atomic volume per length b of dislocation; this is in effect a 'migration volume' of dislocations while the Seeger and Haasen effect (see above) corresponds to a 'formation volume'. The so-called *strength-differential effect* in martensitic steels is explained in the same way (Fletcher *et al.*, 1974): the flow stress is significantly greater in uniaxial compression than in uniaxial tension; this is attributed to the existence of a formation volume for the kink pairs, more energy being expended to form a kink pair against the hydrostatic pressure component of the compression regime than in the tensile regime.

Fig. 5.4. Load-elongation curves for potassium at $T = 77$ K and various pressures (after Chua & Ruoff, 1975).

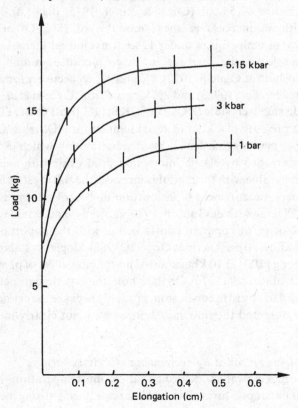

(iii) *Cross-slip*

Whatever the model for cross-slip, it involves some constriction of dissociated dislocations. If a second-order volume expansion is associated to stacking faults (as it is to the core) it is reasonable to expect hydrostatic pressure to favour constriction of dislocations, hence to lower the stress σ_{III} for cross-slip (see fig. 4.3). Indeed, such an effect has been found in NaCl single crystals compressed at room temperatures and at pressures up to 10 kbar (Aladag *et al.*, 1970); the crystals exhibited a relative decrease in σ_{III} (fig. 5.5):

$$\frac{\mathrm{d} \ln \sigma_{III}}{\mathrm{d}P} = -0.1 \text{ kbar}^{-1}$$

Further experiments on polycrystals (grain size $\approx 100 \ \mu$m) confirmed this effect: the flow stress at $\varepsilon = 5\%$ was reduced by 20% (fig. 5.6) and wavy coarse slip lines were observed; the effect of pressurization before the tests was shown to be insignificant, as expected (Auten *et al.*, 1973).

The lowering of σ_{III} with increasing pressure in NaCl is attributed to the

Fig. 5.5. Pressure dependence of the flow stress at the beginning of stage II and III (cross-slip) in NaCl single crystals (at *RT*). Stage II (linear hardening) is independent of *P*. (Experiments on two different crystals.) (After Aladag *et al.*, 1970.)

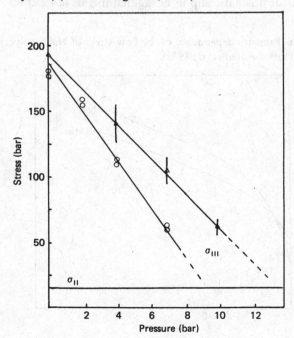

dilatation associated with the stacking faults on {110} planes, predicted by Fontaine (1968). Fontaine & Haasen (1969) calculated the stacking-fault width d as a function of pressure and found that:

$$d \propto \frac{\mu}{\gamma_0 + \alpha P}$$

where μ is the shear modulus, γ_0 the stacking-fault energy at atmospheric pressure and α a constant proportional to the dilatation in the lattice in the vicinity of the stacking fault. The increase in γ with P outweighs the increase in μ and the stacking-fault width decreases with increasing pressure (linearly, in first approximation) which obviously favours cross-slip. The value found for d $\ln \sigma_{III}$/dP is in good agreement with the experiments (Haasen *et al.*, 1970). Belzner & Granzer (1977) calculated the influence of hydrostatic pressure on the energy of the $\frac{1}{2}[110]$ (110) edge dislocation in NaCl, taking into account the discrete atomistic configuration of a finite stacking fault. They found that the dissociation of a dislocation into two partials is not energetically favourable for $P > 2.4$ kbar.

The same effect was found for the metal beryllium (Bedère *et al.*, 1970): the critical resolved shear stress for prismatic slip exhibited a marked decrease from $\sigma = 5.75$ kbar at atmospheric pressure to 5.30 kbar at 10 kbar pressure; this, too, was attributed to the effect of pressure on cross-slip thought to control prismatic slip in hexagonal metals (§ 4.2.2).

Fig. 5.6. Pressure dependence of the flow stress of NaCl polycrystals (at RT) (after Auten *et al.*, 1973).

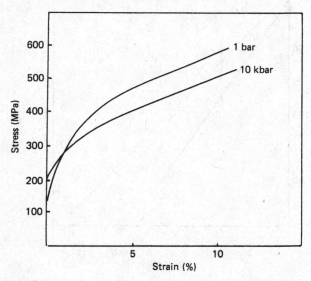

5.2.3 Effect of pressure on hydrolytic weakening in quartz

Natural quartz SiO_2 is usually very strong, indeed almost undeformable under normal laboratory conditions, in a dry environment. However, quartz could be easily deformed at 900–1000 °C, under 15 kbar confining pressure. (Flow stress of 200 MPa under a strain-rate $\dot{\varepsilon} \approx 10^{-5}$/s (Griggs, 1967).) The confining medium was talc, which released water in this temperature range. The same unexpected mechanical weakness was also found in synthetic quartz crystal, hydrothermally grown, which had a high grown-in concentration of hydroxyl ions OH (Griggs & Blacic, 1965; Hobbs *et al.*, 1972; Blacic, 1975). The weakness appeared above a critical temperature which decreased for increasing OH content (measured by infrared spectroscopy). Kekulawala *et al.* (1978) found that natural amethyst with an infrared absorption spectrum similar to that of synthetic 'wet' quartz behaved in similar manner in mechanical tests (fig. 5.7). Conversely, mechanically weak quartz, synthetic or natural, is easily made very strong by heating at atmospheric pressure and precipitating water into bubbles. The weakening effect of water in the lattice was attributed to the

Fig. 5.7. Stress–strain curves for different varieties of quartz ($\dot{\varepsilon} = 10^{-5}$/s, $T = 800$ °C, $P = 300$ MPa). The heat-treated specimens were heated at 900°C for 49 h (wet synthetic quartz) and 27 h (amethyst) (after Kekulawala *et al.*, 1978).

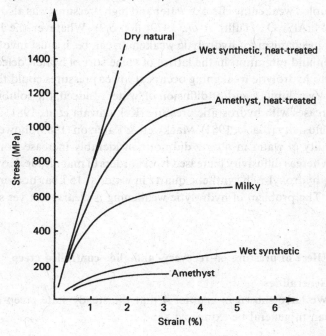

hydrolysis of Si—O—Si bonds, the migration of H_2O replacing the strong covalent bonds by weaker SiOH$\cdots$HOSi hydrogen bonds (Frank–Griggs hypothesis), thus making dislocation glide easier (Griggs, 1967). Transmission electron microscopy of weak quartz samples led McLaren & Retchford (1969) to propose that migration of water to dislocation made climb, hence recovery, easier. Finally, Hobbs *et al.* (1972) and Griggs (1974) adapted to the case of quartz the model for deformation of covalent semiconductor crystals of Alexander & Haasen (1968), assuming that water diffusion to dislocations controlled their multiplication rate. However, what is really the nature of the 'water' that diffuses to dislocations is not clear. The mechanical weakness of quartz seems correlated to the existence of a broad 'gel type' IR absorption band different from the molecular water 'ice' band (Paterson & Kekulawala, 1979). Recently McLaren *et al.* (1983) suggested that the hydroxyl defect responsible for hydrolytic weakening of quartz might be the $(4H)_{Si}$ defect: four protons substituting for one Si^{4+} ion. Pressure was found to play an important role in hydrolytic weakening. Whereas the effect was originally found in natural quartz heated in a wet environment at 15 kbar (Griggs, 1967), it was impossible to reproduce at pressures of 3 to 5 kilobars: natural quartz heated in water at these lower pressures remained very strong (Kekulawala *et al.*, 1978, 1981) but became weak if heated in water at 15 kbar confining pressure (Mackwell & Paterson, 1983) (fig. 5.8).

The coupled weakening effect of water and high pressures was also found in albite $NaAlSi_3O_8$ (Tullis *et al.*, 1979) (fig. 5.9). Whatever the largely unknown mechanism for hydrolytic weakening can be, it must involve the dissolution and migration in the lattice of some sort of 'water' defect; the fact that no hydrolytic weakening occurs at lower pressures could then be due to a low solubility and/or diffusion of 'water'. Indeed, the solubility of water increases with hydrostatic pressure (Kekulawala *et al.*, 1981) and so does its diffusivity (Blacic, 1981). Mackwell & Paterson (1983) showed that the solubility of water in quartz did not considerably increase from 3 to 15 kbar, whereas diffusivity increased by six orders of magnitude. However, annealing hydroxyl-rich synthetic quartz in water at 15 kbar does make it stronger! The problem of hydrolytic weakening is clearly not yet solved.

5.3   Effect of pressure on recovery- and glide-controlled creep

5.3.1   Generalities

We have shown in chapter 4 that the steady-state creep-rate of crystals can in general be expressed as:

$$\dot{\varepsilon} = \dot{\varepsilon}_0 \left(\frac{\sigma}{\mu} \right)^n \exp \left(-\frac{\Delta G}{kT} \right) \tag{5.3}$$

where the preexponential term $(\sigma/\mu)^n$ reflects the stress dependence of the enthalpy of activation of the process controlling the average velocity of the dislocations. The variation of ΔG with hydrostatic pressure is expressed in the *activation volume for the rate-controlling process*:

$$\Delta V = \frac{\partial \Delta G}{\partial P} \Big)_T \tag{5.4}$$

(i) In the case of glide-controlled (or cross-slip-controlled) creep, ΔG is stress-dependent and $n \approx 2$; it is usually possible to derive an expression for ΔG from an elastic model of the activated configuration. Whatever the model may be, it leads to an expression of ΔG proportional to μb^3 or $\mu \Omega$ (Ω: atomic volume). *The activation volume, whose physical basis is some anharmonic expansion of the dislocation core, can then be expressed in terms of the pressure dependence of the shear modulus and the atomic volume.*

Fig. 5.8. Effect of hydrothermal heat treatment under high pressure on the flow stress of natural quartz (after Mackwell & Paterson, 1983) ($P = 300$ MPa, $T = 900\,°C$, $\dot{\varepsilon} = 10^{-5}$/s).

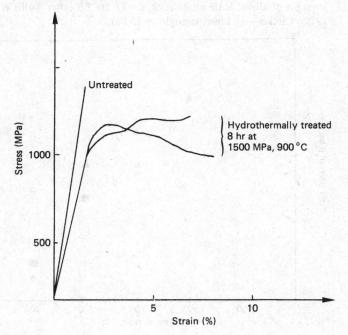

(ii) In the case of diffusion-controlled creep, through climb of dislocations, $n = 3$, and $n = 1$ for diffusion creep by transport of matter (chapter 7); in both cases ΔG is the activation free energy of diffusion of the controlling species. As we will see below, *it is also possible, on the basis of elastic models of vacancies, to find an expression for the activation volume in terms of the pressure dependence of the elastic constants.*

The pressure dependence of the creep-rate is embodied in the *apparent activation volume for creep*:

$$\Delta V_c = -kT \frac{\partial \ln \dot{\varepsilon}}{\partial P} \tag{5.5}$$

with (5.3) and (5.4) we obtain:

$$\Delta V_c = -kT \frac{\partial \ln \dot{\varepsilon}_0}{\partial P} + nkT \frac{\partial \ln \mu}{\partial P} + \Delta V \tag{5.6}$$

If the first two terms on the right-hand side are negligible compared with ΔV, which is usually the case, the apparent activation volume for creep is approximately equal to the activation volume of the rate-controlling process.

Measurements of activation volume for creep are experimentally difficult and so far they have been mostly conducted on materials with low melting

Fig. 5.9. Weakening effect of pressure on high-temperature yield strength of albite, Hale albite rock, $\dot{\varepsilon} = 3 \times 10^{-6}$/s (after Tullis *et al.*, 1979). Circles – 10 kbar; triangles – 15 bar.

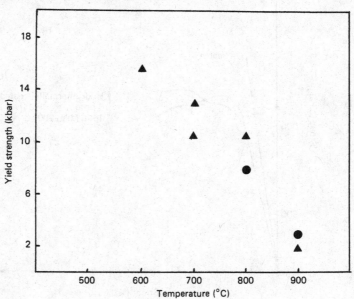

point: lead (Butcher & Ruoff, 1961; De Vries *et al.*, 1963*a*), indium (Chevalier *et al.*, 1967) (fig. 5.10), tin (De Vries *et al.*, 1963*b*), cadmium and zinc (De Vries & Gibbs, 1963). The activation volume for creep is usually close to the atomic volume, hence to the activation volume for diffusion (see Sherby *et al.*, 1970, for tables of data, and Poirier, 1978, for a critical review). The case of aluminium is particularly interesting: McCormick & Ruoff (1969) (fig. 5.11) found $\Delta V = 12.9$ cm^3/mol (≈ 1.3 at. vol.) in good agreement with the value found for self-diffusion (Beyeler & Adda, 1968) in the high-temperature range ($T > 0.6T_m$) where the activation energies for creep and diffusion also coincide. ΔV decreases with temperature. At temperatures between 0.3 and $0.5T_m$ it is equal to about 0.7 at. vol., in the range where cross-slip is thought to control creep-rate, in agreement with the value that can be calculated from the pressure dependence of elastic constants (Poirier, 1978). There are very few measurements of the activation volume for creep in non-metals. Ross *et al.* (1979) measured ΔV for olivine between 1100 °C and 1350 °C for confining pressures up to 15 kbar; they found a mean volume $\Delta V \approx 13.4$ cm^3/mol; although it is comparable to the ionic volume of the oxygen anion (11.6 cm^3/mol) it cannot be used to constrain any physical mechanism for creep in the present state of affairs (see § 4.5.3). The activation volume for creep in caesium chloride CsCl (B2 structure) was measured between 150 °C and 400 °C for confining pressures up to 40 kbar by Heard & Kirby (1981), the value $\Delta G \approx 53$ cm^3/mol

Fig. 5.10. Creep curves of indium at various pressures ($T = 348$ K, $\sigma = 1.8$ bar) (after Chevalier *et al.*, 1967).

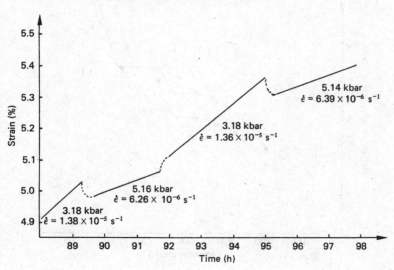

(≈ 1.2 mol. vol.) was found to be compatible with an intrinsic vacancy diffusion mechanism.

5.3.2   Correlations and models for the activation volume

Owing to the scarcity of experimental measurements of activation volumes for diffusion and creep and to their complete absence in the relevant pressure range, estimates of the mantle viscosity (especially for the lower mantle) must rely on estimates of the activation volume at high pressure. Sammis *et al.* (1981) have recently given a clear critical review of the known estimation methods as well as up-to-date tables of experimental results. Note that all the estimations briefly reviewed below refer to activation volumes for diffusion in the intrinsic regime (i.e. for pure crystals at high temperatures).

(i)   *Correlation between ΔV and melting temperature*

For many materials, mostly metals, there is a reasonably good correlation between the activation enthalpy for diffusion and the absolute melting temperature T_m (see Sherby *et al.*, 1977):

Fig. 5.11. Temperature dependence of the activation volume for creep of aluminium at 0 and 3 kbar (after McCormick & Ruoff, 1969). The dashed line indicates the value of the diffusion activation volume from Beyeler & Adda (1968).

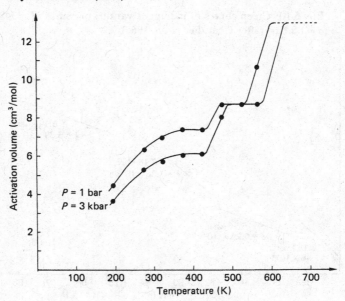

$$\Delta H \approx g R T_\mathrm{m}, \quad \text{with } g \approx 18 \text{ for metals and 30 for silicates}$$

or

$$D = D_0 \exp\left(-\frac{g T_\mathrm{m}}{T}\right) \tag{5.7}$$

Weertman (1970) suggested that the equation (5.7) could be used at all pressures provided that the actual melting point $T_\mathrm{m}(P)$ be used. From eq. (5.7) and the definition of the activation volume comes

$$\Delta V = g R \frac{\mathrm{d} T_\mathrm{m}}{\mathrm{d} P} = \Delta H \frac{\mathrm{d} \ln T_\mathrm{m}}{\mathrm{d} P} \tag{5.8}$$

and with Clapeyron's equation

$$\frac{\mathrm{d} T_\mathrm{m}}{\mathrm{d} P} = \frac{\Delta v_\mathrm{m}}{\Delta s_\mathrm{m}} \tag{5.9}$$

where Δv_m and Δs_m are the variations of volume and entropy at melting

$$\Delta V = g R \frac{\Delta v_\mathrm{m}}{\Delta s_\mathrm{m}} \tag{5.10}$$

Eq. (5.10) justifies the fact that crystals which contract at melting (Pu, Ge, Si, Bi, Sb, Ga, Ice) have a negative activation volume.

Sammis *et al.* (1981) give another expression for (5.8), assuming a linear relation $\Delta H = \Delta H_0 + P \Delta V$, with ΔV independent of P:

$$\Delta V = \frac{\Delta H_0 \dfrac{\mathrm{d} T_\mathrm{m}}{\mathrm{d} P}}{T_\mathrm{m} - P \dfrac{\mathrm{d} T_\mathrm{m}}{\mathrm{d} P}} \tag{5.11}$$

Comparison with experimental data shows a good agreement between experimental and calculated values of ΔV for metals (fig. 5.12); the agreement is not so good for alkali halides – values of ΔV calculated from (5.11) are at least 50% too large (this is confirmed by Heard & Kirby (1981), who find the calculated value for CsCl too large by 57%).

(ii) *Elastic models*

On the basis of simple models for vacancies, Keyes (1958) found that the correlation

$$\Delta V = \frac{4}{B} \Delta H \tag{5.12}$$

where B is the bulk modulus, was reasonably verified by the experimental data then available.

The activation volume can be calculated in terms of the activation free energy by assuming that the internal energy is entirely due to the elastic distortion and dilatation energy associated with the point defects responsible for diffusion (usually vacancies). Keyes (1963) assumed that the energy was only due to shear strain around the defect and found that:

$$\Delta V = \Delta G \left[\left(\frac{d \ln \mu}{dP} \right)_T - \frac{1}{B_T} \right] \tag{5.13}$$

where μ is the shear modulus and B the isothermal bulk modulus. This can be seen simply (if less rigorously) by writing:

$$\Delta G \propto \mu \Omega \tag{5.14}$$

where Ω is the atomic volume. Then:

$$\Delta V = \left(\frac{\partial \Delta G}{\partial P} \right)_T = \Delta G \left(\frac{\partial \ln \Delta G}{\partial P} \right)_T \tag{5.15}$$

Eq. (5.13) follows immediately.

Comparison of values of ΔV calculated by (5.13) with experimental values shows that Keyes' approach leads to a systematic underestimation of ΔV (Sammis *et al.*, 1981); the assumption that the energy is entirely dilatational is better although it leads to a slight overestimation of ΔV.

Fig. 5.12. Correlation between experimental and calculated diffusion activation volume for various crystals. The solid line was calculated with eq. (5.11) (Sammis *et al.*, 1981).

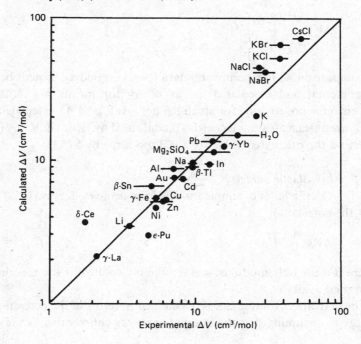

In most cases the observed values of ΔV lie between the shear and dilatational model estimates.

As the activation free energy for diffusion is not always known, it is interesting to find the variation of ΔV with P, purely in terms of the elastic constants and their pressure derivative. O'Connell (1977) assumed that the activation volume for diffusion is approximately equal to the formation volume of vacancies and considered the vacancy as a cavity in the solid under pressure, for which it is possible to define an effective bulk modulus $B_c = (4/9)B$. O'Connell's procedure amounts to expanding the bulk modulus of the cavity to first order in P:

$$-B_c = \frac{\partial P}{\partial \ln V} = -B_c(0)\left[1 + \frac{B'_c(0)}{B_c(0)} P\right] \qquad (5.16)$$

where $B_c(0)$ and $B'_c(0)$ are the bulk modulus of the cavity and its pressure derivative for zero pressure.

Integrating eq. (5.16) gives, in effect, a Murnaghan's equation of state for the cavity (vacancy):

$$\frac{\Delta V(P)}{\Delta V(0)} = \left[1 + \frac{B'_c(0)}{B_c(0)} P\right]^{-1/B'_c(0)} \qquad (5.17)$$

Karato (1981*a*) uses a similar approach by writing that the pressure dependence of ΔV is the same as that of the atomic volume and using directly the Birch equation of state.

It is also interesting to note that the variation of ΔV with P can be very simply found by taking the derivative of ΔV, given by Keyes' relation (5.15), and using the definition of ΔV.

Let us write (5.13) as

$$\Delta V = A\Delta G \qquad (5.18)$$

with

$$A = \frac{\partial \ln \mu}{\partial P} - \frac{1}{B} \qquad (5.19)$$

we have:

$$\frac{\partial \Delta V}{\partial P} = A \frac{\partial \Delta G}{\partial P} + \Delta G \frac{\partial A}{\partial P} = A\Delta V + \Delta V \frac{\partial \ln A}{\partial P}$$

or

$$\frac{\partial \ln \Delta V}{\partial P} = A + \frac{\partial \ln A}{\partial P} \qquad (5.20)$$

We see now that, with all the assumptions of Keyes' model, we have an expression of the derivative of ΔV only in terms of the elastic constants B and μ and their derivatives with respect to P. This expression should obviously be valid for all processes (diffusional or not) for which the activation enthalpy is proportional to the shear modulus (eq. (5.14)). It can be especially valuable for creep in the lower mantle where the only known parameters are the seismically determined elastic constants.

The semi-empirical approach of Karato (1981*b*) requires assumptions regarding many physical quantities such as ionic radius or dielectric constants which are generally unknown, but are reflected in the anharmonic behaviour of elastic constants.

5.3.3   Application to the lower mantle

As soon as it is realized that the activation volume for creep must decrease as pressure increases, it follows that, for reasonable values of the activation energy for creep, the influence of pressure on viscosity cannot counteract the influence of temperature below a certain depth, so that even for flat, near adiabatic, geotherms the viscosity of the lower mantle does not increase much with depth and can even decrease, irrespective of the pressure dependence chosen for ΔV (Sammis *et al.*, 1977; O'Connell, 1977; Karato, 1981*a*).

We will demonstrate it again here by using eqs. (5.19) and (5.20) to calculate the variation of ΔV with depth using the seismically determined actual values of the elastic constants at all depths (Dziewonski *et al.*, 1975, see table 5.1) (fig. 5.13). The ratio $\eta(z)/\eta_0$ of the viscosity at a depth z to that

Fig. 5.13. Depth dependence of the activation volume for creep of the lower mantle (normalized to ΔV_0 at 670 km). $\Delta V(z)$ was calculated using eqs. (5.19) and (5.20) and the seismic elastic constants of Dziewonski *et al.* (1975).

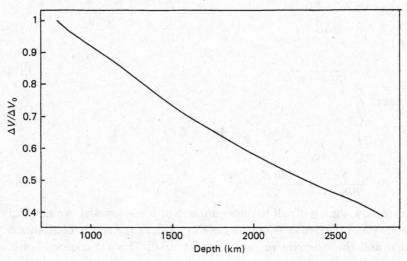

at the top of the lower mantle ($z = 670$ km) is calculated by assuming that the value of the activation enthalpy for creep at 670 km is $\Delta H_0 = 120$ kcal/mol and $\Delta H_0 = 170$ kcal/mol and that ΔV_0 at 670 km is given by Keyes' relation (5.12) (respectively 7 and 10 cm^3/mol). The geotherm $T(z)$ chosen is that of Anderson (1982) (fig. 5.1).

Then:

$$\frac{\eta(z)}{\eta_0} = \exp\left[-\frac{\Delta H_0 + P_0 \Delta V_0}{RT_0}\right] \exp\left[\frac{\Delta H_0 + P\Delta V}{RT}\right] \tag{5.21}$$

We see in fig. 5.14 that, after an increase of not more than an order of magnitude, the viscosity could decrease for most of the extent of the lower mantle and that it could be lower near the core boundary than at 670 km, even if the value of ΔV_0 is higher than given by (5.12): for $\Delta H_0 = 120$ kcal/mol and $\Delta V_0 = 10$ cm^3/mol, $\eta_{\text{CMB}}/\eta_0 \approx 3 \times 10^{-2}$; by contrast the same value of ΔH_0 and a constant value of $\Delta V_0 = 10$ cm^3/mol would give $\eta_{\text{CMB}}/\eta_0 \approx 2 \times 10^{13}$.

As long as we do not know the values of the relevant activation parameters for the lower mantle, the variation of its viscosity with depth will remain unknown. However, we can certainly conclude that the idea of a constant-viscosity lower mantle is perfectly reasonable.

Fig. 5.14. Possible variation of the viscosity of the lower mantle with depth (viscosity normalized to η_0 at 670 km). Calculated using the variation of the activation volume given in fig. 5.13 and various values of the activation enthalpy ΔH_0 and activation volume ΔV_0 at 670 km. 1. $\Delta H_0 = 120$ kcal/mol, $\Delta V_0 = 7$ cm^3/mol. 2. $\Delta H_0 = 170$ kcal/mol, $\Delta V_0 = 10$ cm^3/mol. 3. $\Delta H_0 = 120$ kcal/mol, $\Delta V_0 = 10$ cm^3/mol. 4. $\Delta H_0 = 120$ kcal/mol, $\Delta V(z) = 10$ cm^3/mol.

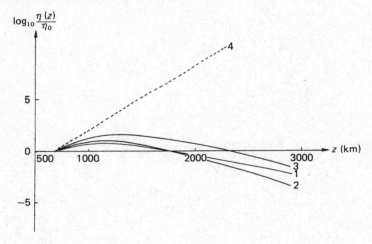

Readings

J. Verhoogen, F. J. Turner, L. E. Weiss, C. Wahrhaftig & W. S. Fyfe, *The Earth*, Holt, Rinehart & Winston (1970), 748 pp. A remarkable introduction to modern geology and geophysics, from minerals to geodynamics.

S. Uyeda, *The new view of the Earth*, W. H. Freeman (1978), 277 pp. A short and clear overview of the global tectonics world.

6

Creep polygonization and dynamic recrystallization

The most important microstructural evolution during creep consists in the formation of misoriented subgrains (polygonization) separated by dislocation walls. The walls result from the rearrangement of the geometrically necessary dislocations that accommodate the plastic incompatibilities between grains or between single crystalline sample and platens. The subgrain structure is in a state of dynamic evolution, walls being created, migrating under stress and being destroyed; the misorientation of the walls increases with strain until a recrystallized grain structure is created by rotation, without migration. For higher stresses and temperatures, the driving force for migration of the boundaries, as well as their mobility, increases and the boundaries may migrate. The subgrain size and the recrystallized grain size both depend only on the applied stress and decrease as the stress increases. Empirical relations between grain size or subgrain size and stress are found experimentally and are used to estimate the stress responsible for the natural deformation of rocks. However, the view that the subgrain or grain sizes are in equilibrium at a given stress is unfounded. The subgrain size is not an independent variable and has no real effect on the creep-rate unless it can be independently fixed. Grain refinement due to dynamic recrystallization does not seem sufficient to cause a change of creep mechanism from power-law to diffusion creep.

6.1   Generalities

The most conspicuous structural feature of high-temperature recovery creep of metals, ceramics and minerals, consists in the 'fragmentation' of crystals into 'subgrains'. Subgrains are blocks of constant crystalline orientation, containing few dislocations, separated by dislocation walls (Jenkins & Mellor, 1935; Wyon & Crussard, 1951; McLean, 1952). The misorientation between neighbouring subgrains can vary from a few seconds of arc, for loose dislocation walls, to about 10° (see § 2.4.2). There is a hierarchy of scales of subgrains, each subgrain being usually divided in smaller subgrains of lesser misorientation. This phenomenon is known as *creep polygonization*.

Subgrain boundaries form during primary creep; after a few per cent strain, they are usually fully developed and constitute a stable steady-state substructure. The substructure is in fact in a state of dynamic evolution, sub-boundaries being formed and destroyed; however, the mean subgrain size (at all scales) remains constant as strain increases; it varies roughly as the inverse of the applied stress and does not usually depend on temperature.

The misorientation usually increases with strain and it may reach such high values ($\theta > 15°$) that it becomes more proper to talk of grains than of subgrains. For higher temperatures and applied stress, the high-angle boundaries may migrate, giving rise to a new recrystallized grain structure. The phenomenon of modification of the grain structure during creep is known as *dynamic recrystallization*.

6.2   Creep polygonization

6.2.1   Observation techniques

Polygonization can be revealed by several techniques based on the observation either of the subgrain boundaries or of the misorientation between subgrains.

(i) *Observation of the boundaries*

The boundaries of the finer-scale substructure can be directly observed by transmission electron microscopy, a technique which allows a precise characterization of the boundaries and of the dislocations that compose them (e.g. Bretheau *et al.*, 1981, for Cu_2O; Caillard & Martin, 1982, for aluminium) (fig. 6.1). However, this technique is not adequate for investigating the coarser substructure, and it must be complemented by other techniques. Boundaries can be revealed at the surface by etch pits, obtained by dipping the samples in suitable reagents (Robinson, 1968; Mendelson, 1961; Wegner & Christie, 1974, 1983) (fig. 6.2). In transparent crystals, methods can sometimes be found for decorating the boundaries; in olivine, for instance, heating in air at 900 °C causes iron oxide particles to precipitate on dislocations making them visible in optical microscopy (Kohlstedt *et al.*, 1976) (fig. 6.3). Boundaries can also be seen and analysed, using non-destructive X-ray topographic methods (Weissman & Kalman, 1969); the reflection Berg–Barrett method, in particular, is often used on massive single crystalline creep samples (fig. 6.4).

(ii) *Observation of the misorientation*

Misoriented subgrains exhibiting differences in reflectivity can often be seen at the surface of creep samples, with the naked eye or a low-power microscope.

Transparent birefringent (non-cubic) crystals can be examined in 30-μm-thick petrographic thin sections, in polarized light; the misoriented subgrains exhibit various degrees of extinction and can be clearly seen (fig. 6.5).

Polarized light can also be used in reflection in the case of opaque metals with transparent birefringent coatings, like aluminium or magnesium (Ion *et al.*, 1982). Subgrains can be seen by orientation contrast in X-ray topograms. Creep substructures in electronically conducting crystals are clearly revealed in electronic images obtained with a scanning electron microscope or an electron microprobe; the crystalline contrast results from intensity variations in the anomalous absorption when the subgrain orientation varies near the Bragg angle (Maurice *et al.*, 1973) (fig. 6.6).

6.2.2 Formation and evolution of the substructure

Even though creep polygonization practically always accompanies high-temperature recovery creep, its existence does not result

Fig. 6.1. Subgrain boundaries in olivine (Pallasite meteorite 'Imilac'). Transmission electron microscopy (scale bar: 1 μm) (courtesy A. M. Marie).

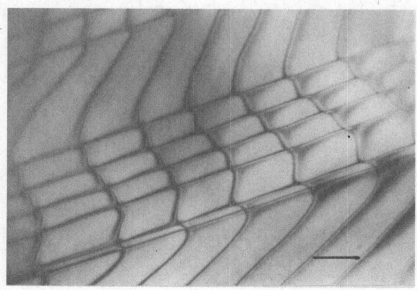

from any intrinsic necessity but can be seen as contingent. In fact, the subgrain boundaries constitute a lower-energy configuration for the population of excess dislocations of one sign needed to accommodate locally the strain gradients (or the incompatibility) when the deformation is not uniform. This is always the case for polycrystals, each grain deforming on differently oriented slip systems; internal stresses arise near grain boundaries and can be relaxed at high temperatures by excess dislocations or '*geometrically necessary*' *dislocations* (Ashby, 1970). For the same reason, the local curvature of the glide planes (see §1.1.3) can be taken up by dislocations of one sign (Nye, 1953) which evolve by glide and climb toward a lower-energy configuration of parallel tilt boundaries. The boundary

Fig. 6.2. Subgrain boundaries in NaCl, deformed at 680 °C, $\sigma = 35$ g/mm^2. Etch pits.

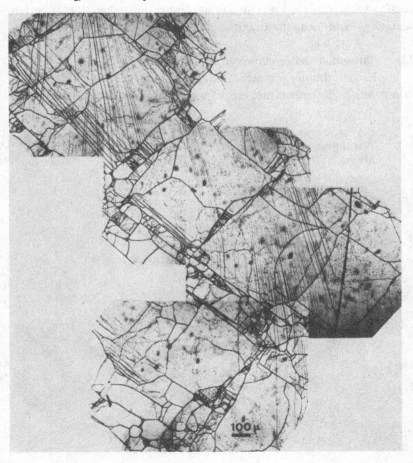

conditions imposed by the neighbouring grains also often give rise to rotations of the slip planes in opposite senses and create misoriented regions separated by diffuse boundaries (deformation bands). It is therefore not surprising that the creep substructure is finer in the regions of the grain boundaries where the strain gradient is larger: the grains in a polycrystal often exhibit a 'core and mantle' structure (Gifkins, 1976). Single crystals deformed in creep also exhibit polygonization in most cases; this is due to the fact that they are deformed between hard unyielding platens with no lubrication so that the glide planes must rotate and acquire a curvature (§ 1.1.3). In rare cases, where the lubrication is good and the single crystal can deform in a uniform fashion, no creep substructure appears, even for large strains: this was the case for a single crystal of olivine deformed by Durham *et al.* (1977). Most studies on the formation of the substructure have been conducted on single crystals, by etch-pit techniques: the principal results were reviewed by Takeuchi & Argon (1976*a*). The process of formation of the substructure is identical in most cubic crystals, whether they are metals like copper (Hasegawa *et al.*, 1971) or molybdenum (Clauer *et al.*, 1970), oxides like MgO (Hüther & Reppich, 1973) or $MgAl_2O_4$ spinel (Doukhan *et al.*, 1973), or halides like NaCl (Poirier, 1972) or AgCl (Pontikis & Poirier, 1975). In a first stage, during primary creep, parallel

Fig. 6.3. Subgrain boundaries in olivine (xenolith from San Quintin, Baja California, Mexico). Decoration by oxidation. Tilt boundaries in plane (100) (scale bar: 100 μm) (courtesy J. C. Mercier).

Fig. 6.4. Subgrain boundaries in deformed spinel, Berg–Barrett technique (scale bar = 0.1 mm) (courtesy R. Duclos).

Fig. 6.5. Subgrains in naturally deformed olivine (xenolith from Black Rock Summit, Nevada), optical microscopy, plane-polarized light (scale bar = 1 mm) (courtesy J. C. Mercier).

Fig. 6.6. Subgrains in gold single crystal deformed by 22% at., $T =$ 840 °C and $\sigma = 200$ g/mm². Electronic image (scale bar = 1 mm).

dislocation tilt walls form, normal to the primary slip direction, accompanied by twist walls parallel to the slip planes; as the strain increases, the dislocation content of the walls (hence their misorientation) increases. If a secondary slip system comes into play, as is the case for many cubic crystals tested in conditions of duplex slip, new dislocation walls eventually subdivide the elongate cells into smaller equiaxed subgrains. The resultant steady-state structure is more or less equiaxed, but not always uniform. If there is only one predominant slip system, as in olivine (Durham et al., 1977) or rutile (Hirthe & Brittain, 1963), the characteristic creep substructure consists of parallel tilt subgrain boundaries.

The mean steady-state coarse subgrain size (or tilt boundary spacing) d, does not depend on temperature; it characteristically decreases as the applied stress increases according to the empirical law:

$$\frac{d}{b} = K\left(\frac{\sigma}{\mu}\right)^{-p} \tag{6.1}$$

where p has a value close to 1 or slightly smaller ($0.7 \leqslant p \leqslant 1$); the constant K has a value of about 10 for metals and 25 to 80 for ionic crystals and oxides (Takeuchi & Argon, 1976a) (fig. 6.7). We may notice that *for eq. (6.1) to be dimensionally correct (and for K to be dimensionless), p must be equal to 1 (which it is in most cases)*. The inverse relationship between subgrain size and stress has been used to estimate the stress to which naturally deformed samples were subjected; we will discuss this procedure in §6.4 but we may already point out here that the coarse subgrain size alone may reflect the stress state during creep. The small-scale microstructure inside the coarse grains consists of free dislocations (with a density proportional to σ^2) and dislocation walls that can be seen in TEM to form, migrate and be destroyed (Caillard & Martin, 1982). Transmission electron microscopy of specimens previously deformed in creep provides an empirical relation of the same type as (6.1). However, the dislocation microstructure inside the coarse subgrains is very labile and responds rapidly to any changes in the stress conditions; it is therefore quite possible that the cell structure observed in TEM after creep reflects only the stress-free anneal inevitable at the end of the experiments, during which dislocations may rearrange into a lower-energy configuration. In the same manner, no undue emphasis should be put on the observation, often made in TEM, that the dislocation wall misorientation increases only slightly with strain (Takeuchi & Argon, 1976a); it may reflect nothing more than the post-creep rearrangement of free dislocations whose density increases with strain.

On the other hand, the coarse subgrain walls keep a longer memory of the creep conditions, even though they constantly evolve and migrate during creep. The coarse subgrain boundaries have been observed to migrate during creep in aluminium (Exell & Warrington, 1972) and sodium chloride (Guillopé & Poirier, 1979), contributing 10 to 20% of the total creep strain (by what is, in effect, collective dislocation slip). In many cases, the subgrain boundaries were observed to increase in misorientation with strain up to large values (becoming in fact high-angle grain boundaries in some cases). Progressive misorientation of the boundaries was thus reported in aluminium (McLean, 1952), nickel (Richardson *et al.*, 1966) and quartz (Hobbs, 1968); evidence for progressive misorientation in naturally deformed minerals was given in the case of quartz (White, 1973) and olivine (Poirier & Nicolas, 1975).

6.2.3   Subgrain size and creep-rate

The problem of the existence of a dependence of creep-rate on subgrain size has been the object of much debate, although it does in fact

Fig. 6.7. Correlation between normalized subgrain size d/b and applied stress τ/μ, for single crystals of various structures (Takeuchi & Argon, 1976a).

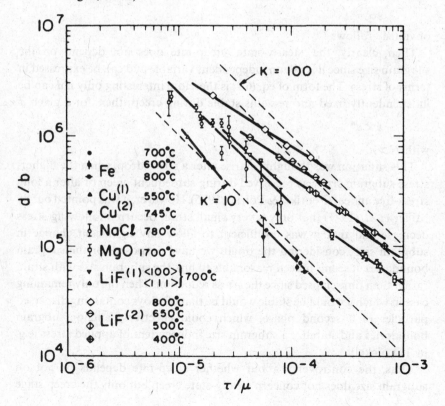

boil down to a problem of terminology. The case for a subgrain-size dependence of the creep-rate (Sherby *et al.*, 1977) is built around the observation that the stress exponent of the creep-rate determined by a stress-drop technique is higher than the one determined by the slope of the log $\dot{\varepsilon}$–log σ plot; in other words, the constant-structure exponent $N = 1/m$ is higher than the 'steady-state' exponent n (see §1.2.3). As the most obvious structural feature of creep is the subgrain size d, it is reasonable to assume that the subgrain size is the structure parameter and that 'creep at constant structure' means 'creep at constant subgrain size'. It has indeed been verified, by etch-pit measurements on silver chloride, that the subgrain size following a stress drop remains equal to the subgrain size given by eq. (6.1) with $p = 1$ for the higher stress (Pontikis & Poirier, 1975). The creep rate may be written:

$$\dot{\varepsilon} = K d^{N-n} \sigma^N \qquad (6.2)$$

For 'steady-state' creep:

$$d \propto \sigma^{-1}$$

The classical power-law dependence:

$$\dot{\varepsilon} \propto \sigma^n$$

obviously follows.

Then, clearly, the 'steady-state' creep-rate does not depend on the subgrain size since d is not an independent variable and can be expressed in terms of stress. The form of eq. (6.2) is therefore interesting only if d can be independently fixed and remains stable during creep; then, for a given d

$$\dot{\varepsilon} \propto \sigma^n$$

with $N > n$.

This situation was thought to arise after a stress drop, when the higher-stress subgrain size was preserved during subsequent creep or after a long stress-free anneal (Pontikis & Poirier, 1975). However, it was pointed out by Miller *et al.* (1977) that only a very small strain occurred following stress decrease and that it was insufficient to allow any significant change in subgrain size; considering the origin we have proposed for the subgrain boundaries, it seems indeed reasonable to link the subgrain size with strain rather than time elapsed since the stress reduction. Then the only remaining case in which d would be stable would be that of alloys containing dispersed particles of a second phase, which could effectively anchor subgrain boundaries and stabilize a subgrain size independent of applied stress (e.g. in TD nickel).

Thus, the controversy about whether creep-rate depends or not on subgrain size does not concern steady-state creep but only the creep stage following immediately a stress drop, which is interpreted in different ways (Parker & Wilshire, 1976; Miller *et al.*, 1977; Sherby *et al.*, 1977). This is probably not a major problem.

6.3   Dynamic recrystallization

6.3.1   Recrystallization processes

We will define dynamic recrystallization as a *deformation-induced reworking of the grain sizes, shapes or orientations with little or no chemical change* (Poirier & Guillopé, 1979). As the change of grain structure was usually observed after high-temperature deformation in metals that also easily recrystallized statically (see §2.3.2), the existence of recrystallization during creep or constant strain-rate deformation was denied for a long time. It is now well accepted, and its occurrence in metals, minerals and organic crystals can be followed optically *in situ* (Urai *et al.*, 1980; Means, 1983) or through its effects on the creep curves or stress–strain curves (figs. 6.8, 6.9, 6.10).

Dynamic recrystallization and dynamic recovery were long seen as competing phenomena in metals (Hardwick *et al.*, 1961). Indeed, this view was reasonable as long as the only process for dynamic recrystallization was thought to be nucleation and growth of new grains by grain-boundary

Fig. 6.8. Effect of dynamic recrystallization on creep curves of pure nickel polycrystals ($T = 965$ °C) (after Richardson *et al.*, 1966).

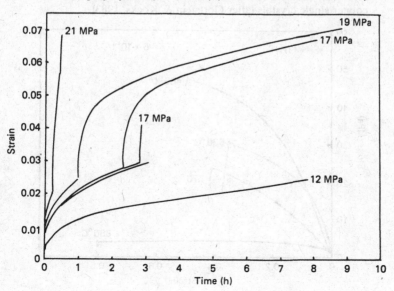

migration; the driving force for grain-boundary migration being due to the difference in strain energy between deformed grains and dislocation-free nuclei (§2.4.5), it obviously decreases by a large amount when the dislocations in the deformed grains rearrange into lower-energy configura-

Fig. 6.9. Effect of dynamic recrystallization on stress–strain curves of pure nickel polycrystals ($T=934$ °C) (after Luton & Sellars, 1969).

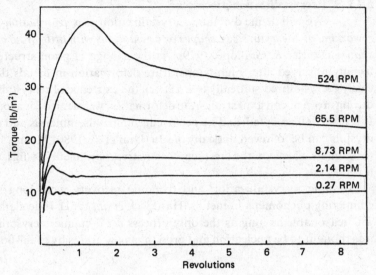

Fig. 6.10. Effect of dynamic recrystallization on stress–strain curves of copper single crystals (after Gottstein & Kocks, 1983).

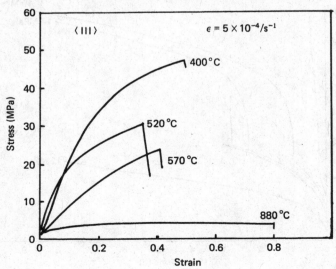

tions by polygonization. However, it seems now clear that polygonization, far from impeding recrystallization, may further it, as we will see presently.

The grain-structure modification in polycrystals or single crystal may be achieved in two ways, which we will call 'rotation recrystallization' and 'migration recrystallization' (Poirier & Guillopé, 1979).

(i) *Rotation recrystallization*

This results from progressive misorientation, without appreciable migration, of the subgrain boundaries formed during creep polygonization: the grains 'fragment' into subgrains which eventually become so misoriented that they truly become individual grains (fig. 6.11); however, they usually retain some orientation kinship that allows the reconstitution of the former parent grain (Poirier & Nicolas, 1975; Urai, 1983). This type of recrystallization has been observed in many materials: quartz (Hobbs, 1968; White, 1973), calcite (Schmid *et al.*, 1980), $NaNO_3$ (Tungatt & Humphreys, 1981), NaCl (Guillopé & Poirier, 1979), olivine (Poirier & Nicolas, 1975; Karato *et al.*, 1980), bischofite $MgCl_2, 6H_2O$ (Urai, 1983); it has also been observed in hot worked aluminium (Perdrix *et al.*, 1981), a metal that was supposed to polygonize and not recrystallize owing to its high stacking-fault energy. Rotation recrystallization is clearly not competitive with polygonization, indeed it necessarily results from it at

Fig. 6.11. Rotation recrystallization in NaCl single crystal deformed by 70% at $T = 600$ °C and $\sigma = 40$ g/mm^2. Etch pits. Subgrains are labelled with their misorientation in degrees with respect to the subgrain 0 (near the centre) 30 denotes a 30° rotation in the counter-clockwise sense and 30 a 30° rotation in the clockwise sense) (scale bar = 1 mm) (courtesy M. Guillopé).

large strains, if the incompatibilities are maintained. The boundaries can migrate slowly under stress (§ 6.2.2) in the low-velocity regime characteristic of impurity-loaded boundaries (§ 2.4.6).

(ii) *Migration recrystallization*

This results from the rapid migration of grain boundaries between dislocation-free nuclei and deformed grains (fig. 6.12), exactly as in static primary recrystallization (see § 2.4.7). In fact, there is no evidence for the existence of nuclei arising from statistical fluctuations in the classical sense of the nucleation theory; rather, the 'nuclei' seem to be constituted by dislocation-free regions of subgrains (or whole small subgrains) bounded by a mobile stretch of high-angle boundary that can bulge out into a deformed neighbouring grain by strain-induced boundary migration (SIBM). The driving force, as in static recrystallization, arises from the strain-energy difference between deformed and virgin grains. The strain energy in a volume is proportional to the dislocation density ρ, and to the area of subgrain and grain boundaries (of specific energy γ_{sgb} and γ_{gb}) it contains. In the deforming grain, the dislocation density is usually proportional to σ^2 and the area of subgrain and grain boundaries per unit volume is inversely proportional to the subgrain and grain sizes, which in turn decrease with

Fig. 6.12. Migration recrystallization in NaCl single crystal. The vertical grain boundary (left) moves towards the left, consuming the heavily polygonized grain at left. The new grain (right) is being polygonized in turn (scale bar = 0.1 mm) (courtesy M. Guillopé).

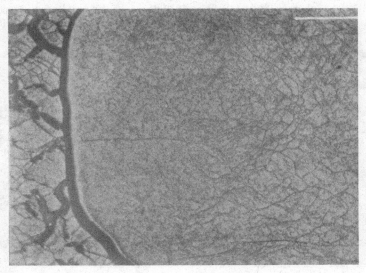

increasing stress (§ 6.2.2 and § 6.3.3). The result is, that the driving force F for a grain boundary between a region of deformed grains of size D_1, containing subgrains of size d_1 and a dislocation density ρ, and a recrystallized region of grain size $D_2 \propto \sigma^{-1.2}$ can be written (Poirier & Guillopé, 1979):

$$F \approx 3\left[\gamma_{gb}\left(\frac{1}{D_1} - A\sigma^{1.2}\right) + B\gamma_{sgb}\sigma\right] + C\rho\sigma^2$$

The driving force is an increasing function of the applied stress. For high driving forces and high intrinsic mobilities (i.e. at high stress and temperature), the grain boundaries can free themselves from the pinning impurity atoms (§ 2.4.6) and move at high velocity, consuming the deformed material. Obviously, in the case of dynamic recrystallization, the newly formed grains behind the moving boundary get deformed and polygonized; this eventually lowers the driving force, leading to an exhaustion of the recrystallization process or to recrystallization waves, depending on the interplay between the kinetics of recrystallization and deformation (Sellars, 1978).

For a given impurity concentration, there should therefore exist a curve in the stress–temperature plane between the domain where the grain boundaries cannot move rapidly and only rotation recrystallization is possible (low σ and T) and the domain where the migration recrystallization can occur (high σ and T). Such a curve was determined for NaCl (Guillopé & Poirier, 1979) and $NaNO_3$, an analogue of calcite (Tungatt & Humphreys, 1981) (fig. 6.13(b)). Two regimes of velocity for boundary migration were also observed in bischofite[†] (Urai, 1983) and dynamic recrystallization in this material results from a complex interplay between rotation and migration recrystallization.

From all the known examples of these processes, it is seen that rotation and migration recrystallization are not exclusive and that in most cases rotation recrystallization first occurs, a high-angle boundary structure is then built up, sometimes as a 'core and mantle' structure in the grains. For strain larger than a critical strain (about 40 to 60% in most cases), the driving force and the intrinsic mobility (linked to the misorientation angle) of the boundary may be high enough (if the temperature is also high enough) and the migration recrystallization may start. It often starts in the 'mantle' region close to grain boundaries and the small subgrains may be seen as 'nuclei', enlarging by SIBM; most often, the resulting structure may be equiaxed due to the impingement of neighbouring grains (e.g. in

---

[†] The high-velocity regime corresponds to a mechanism not found in metals or most minerals: the strained grains dissolve in a water film at the boundary and the new crystals grow by redeposition of the dissolved material.

Fig. 6.13. Boundary between rotation and migration recrystallization domains. (*a*) NaCl deformed in creep (σ, T plane). Circles: pure NaCl. Triangles: NaCl doped with 500 ppm Sr^{2+}. Squares: natural halite from Dieuze mines. Empty symbols: rotation recrystallization. Closed symbols: migration recrystallization. (After Guillopé, 1981.) (*b*) NaNO$_3$ deformed at constant $\dot{\varepsilon}$ (after Tungatt & Humphreys, 1981).

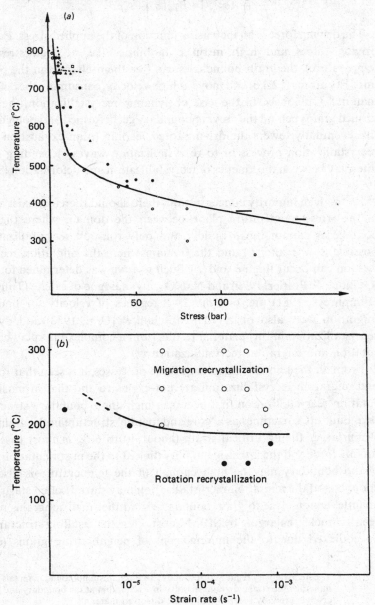

magnesium, 0.8% aluminium, alloy; Ion *et al.*, 1982). The impurity content of the crystals is obviously an important factor and it is seen that the limiting curve between rotation and migration recrystallization moves toward higher stresses and temperatures for impure crystals (Guillopé & Poirier, 1979) (fig. 6.13(*a*)).

6.3.2 Recrystallized grain size

As polygonized subgrains, dynamically recrystallized grains form a continuously evolving structure, characterized by a mean grain size D, which depends only on the applied stress:

$$\frac{D}{b} = K' \left(\frac{\sigma}{\mu}\right)^{-r}$$

(6.3)

Eq. (6.3) is similar to eq. (6.1) for subgrains, but the value of the exponent r is usually somewhat larger than 1 and close to $r \approx 1.2$ (table 6.1) (fig. 6.14).

The size of the recrystallized grains does not usually depend on temperature; however, Ross *et al.* (1980) found that they could fit better the data from their experiments on dunite (olivine) under pressure, with dehydrating talc as a confining medium, by adopting a weak temperature dependence ($Q \approx 14$ kcal/mol); the temperature dependence disappeared when they used a dry confining medium.

In most cases, the grain size is given without mentioning whether it results from rotation or migration recrystallization. In the cases where the two regimes have been clearly observed and the corresponding grain sizes measured, D is larger for migration than for rotation recrystallization (Guillopé & Poirier, 1979; Tungatt & Humphreys, 1981) (fig. 6.15).

We must realize here an important difference between the equation for subgrains (6.3), which can be written

Table 6.1. *Correlation between recrystallized grain size and applied stress*, $D/b = K(\sigma/\mu)^{-r}$

Material	K	r	Reference
Quartzite (wet)	2.65	1.40	Mercier *et al.* (1977)
Quartzite (wet)	231	0.90	Christie *et al.* (1980)
Olivine (dry)	38	1.18	Karato *et al.* (1980)
Dunite (dry)	62	1.27	Ross *et al.* (1980)
Dunite	69	1.23	Mercier *et al.* (1977)
Pyrite	22	1.11	Cox *et al.* (1981)
Halite (rotation)	13.3	1.18	Guillopé & Poirier (1979)
Halite (migration)	178	1.28	Guillopé & Poirier (1979)
Nickel	2×10^4	1.33	Luton & Sellars (1969)

$$\frac{d}{b} = K \left(\frac{\sigma}{\mu} \right)^{-1} \tag{6.4}$$

and the equation for recrystallized grains, which can be written:

$$\frac{D}{b} = K' \left(\frac{\sigma}{\mu} \right)^{-1.2} \tag{6.5}$$

Eq. (6.4) is easily understood physically: even if the details of the process leading to a particular value of K are still obscure, the form of the equation is satisfying: a characteristic dimension of the dislocation structure is inversely proportional to a stress (here, it must be assumed that the applied stress is proportional to the internal stress, but we have seen it is reasonable, most of the time). Such is not the case for eq. (6.5); the experimental evidence definitely rules out the possibility of disregarding the fact that the exponent is not equal to 1; even if the equation is artificially made dimensionally correct by normalizing D and σ, there remains that it has no obvious physical sense and that it must be considered as an empirical relation. So

Fig. 6.14. Relation between recrystallized grain size and applied stress for various materials (see table 6.1). 1 – Wet quartzite (Mercier *et al.*, 1977). 2 – Halite, rotation (Guillopé & Poirier, 1979). 3 – Pyrite (Cox *et al.*, 1981). 4 – Olivine (Karato *et al.*, 1980). 5 – Dunite (Mercier *et al.*, 1977). 6 – Halite, migration (Guillopé & Poirier, 1979). 7 – Wet quartzite (Christie *et al.*, 1980).

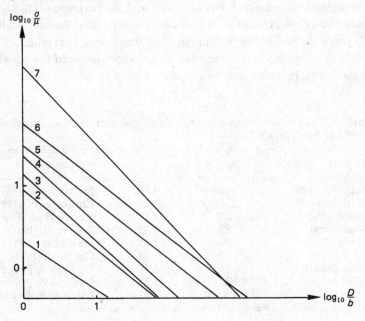

Fig. 6.15. (*a*) Relation between recrystallized grain size and applied stress for NaCl: subgrains – $\sigma/\mu = 20\, b/d$; grains (rotation) – $\sigma/\mu = 9(b/D)^{0.85}$; grains (migration) – $\sigma/\mu = 57(d/D)^{0.78}$ (after Guillopé, 1981). (*b*) Relation between recrystallized grain size and temperature for NaNO₃ (after Tungatt & Humphreys, 1981).

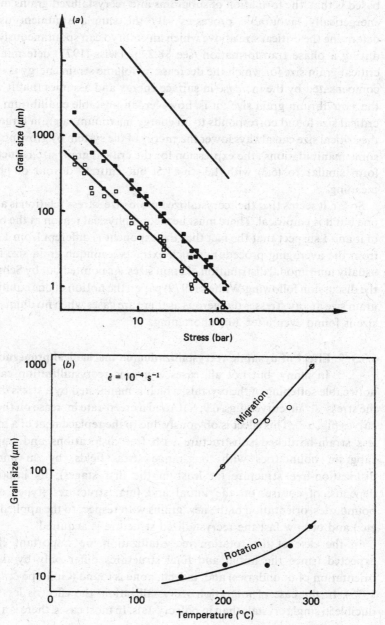

far, all the attempts to justify 'theoretically' the grain-size–stress relation have been remarkably unconvincing. The popular and widely quoted 'theoretical' expression derived by Twiss (1977) does in fact rest on extremely shaky foundations: the fundamental assumption on which it is based is that 'the formation of subgrains and recrystallized grains must be energetically favourable processes'. By the simple argument used to determine the critical size above which an embryo can spontaneously grow during a phase transformation (see §8.2.2), Twiss (1977) determines the critical grain size for which the decrease in volume strain energy is exactly compensated by the increase in surface energy and assumes that it will be the 'equilibrium grain size'; it is, however, an unstable equilibrium as the critical size found corresponds to an energy maximum: a grain larger than the critical size can always lower the energy of the system by growing. 'After some manipulations', the expression for the critical grain size reduces to a form similar to (6.3) with $1.3 < r < 1.5$, but entirely devoid of physical meaning.

So far, it seems that the recrystallized grain-size–stress relation is a useful one but it is empirical. There must be a good physical reason at the bottom of it and I suspect that the fact that the exponent is different from 1 comes from the averaging procedure which extracts a unique grain size from a usually multimodal distribution of grain sizes. As pointed out by Schmid, in the discussion following White's (1979) paper, the notion of one equilibrium grain size at any stress is dangerous as there are cases when no unique grain size is found even after 30% shortening.

6.3.3   Effect of dynamic recrystallization on mechanical properties

In many, but not all, cases, dynamic recrystallization causes a noticeable softening of the crystals, which is manifested by a stress drop on the stress–strain curves (figs. 6.9, 6.10) and a creep-rate increase on the creep curves (fig. 6.8). This effect is obviously due to the replacement of a more or less strain-hardened substructure with free dislocations and imperfect subgrain boundaries with long-range stress fields, by an annealed, dislocation-free structure (at least in the first stages); its importance depends, of course, on the initial and final structures (types of subboundaries, orientation of the new grains with respect to the applied stress, etc.) and of how fast the recrystallized structure is acquired.

In the case of the rotation recrystallization, no important effect is expected (since the initial and final structures differ only by the misorientation of boundaries) and, indeed, none is found (Guillopé & Poirier, 1979). In the case of migration recrystallization, the effect is less reproducible in single crystals than in polycrystals. In most cases, there is a strain-rate increase in creep tests, when recrystallization has affected enough volume in the sample, but the effect is usually more brutal in pure crystals than in impure ones (Richardson *et al.*, 1966). In constant strain-rate tests (most often torsion tests, because they allow large strains, up to the critical strain, to be reached), the stress–strain curves exhibit a stress drop (e.g. in nickel, Gottstein & Kocks, 1983, or in magnesium, Ion *et al.*, 1982) or periodic oscillations (Rossard & Blain, 1958; Glover & Sellars, 1973; Luton & Sellars, 1969). The oscillations usually occur only at large values of the applied stress (or of the Zener–Hollomon parameter); they reflect the occurrence of dynamical recrystallization waves that subside after a while when the straining is fast enough, to start again when the driving force has regained sufficient strength (Jonas *et al.*, 1969; Sellars, 1978).

The softening accompanying dynamic recrystallization may be a cause of plastic instability (§ 1.3) and may give rise to shear zones or ductile faulting, e.g. in magnesium (Ion *et al.*, 1982) or olivine (Post, 1977).

Taking into account the fact that dynamic recrystallization may cause the grain size to decrease if the stress is high enough, it has been suggested (White, 1976; Twiss, 1976) that the creep mechanism of minerals in geophysical settings leading to recrystallization might change from power-law creep to grain-size-dependent creep mechanism, active in fine-grained materials (see chapter 7). This view was criticized by Zeuch (1983), in the case of olivine, on the grounds that the activation energy for the creep of recrystallized olivine (Karato *et al.*, 1982) was higher than the one predicted for diffusion creep. Although the argument does not lack validity, the experimental basis is rather weak in this case, since Karato *et al.* (1982) measured an apparent activation energy for creep, on samples undergoing dynamic recrystallization (i.e. using an average value of a fluctuating creep-rate), thus clearly finding values devoid of any physical significance. Etheridge & Wilkie (1979), more convincingly, defended the thesis that grain refinement is never sufficient to take the creep process out of the power-law field into the diffusion-creep field; we will give here a brief analysis similar in principle to theirs: let us express the viscosity of power-law creep and of diffusion creep by transport of matter by bulk diffusion (Nabarro–Herring creep) or by grain-boundary diffusion (Coble creep) (see chapter 7).

The viscosity of power-law creep is stress-dependent and grain-size-independent:

$$\eta_{\text{PLC}} = K_1 \left(\frac{\sigma}{\mu}\right)^{-2} \tag{6.6}$$

The viscosity of diffusion creep is stress-independent (Newtonian) and grain-size-dependent:

$$\eta_{\text{NHC}} = K_2 \left(\frac{D}{b}\right)^2 \tag{6.7}$$

$$\eta_{\text{CC}} = K_3 \left(\frac{D}{b}\right)^3 \tag{6.8}$$

Eliminating η from (6.6), (6.7) and (6.8), we find the equations of the linear boundaries between the fields of the various creep mechanisms in the log σ–log D plane:

$$\log\left(\frac{D}{b}\right) = \frac{1}{2}\log\left(\frac{K_1}{K_2}\right) - \log\left(\frac{\sigma}{\mu}\right) \tag{6.9}$$

and

$$\log\left(\frac{D}{b}\right) = \frac{1}{3}\log\frac{K_1}{K_2} - \frac{2}{3}\log\left(\frac{\sigma}{\mu}\right) \tag{6.10}$$

Now, the final recrystallized grain size and the applied stress are related by eq. (6.3) with $r \approx 1.2$, hence:

$$\log\left(\frac{D}{b}\right) = \log K_4 - 1.2 \log\left(\frac{\sigma}{\mu}\right) \tag{6.11}$$

It turns out that, in most cases, the values of the constants, K_1, K_2, K_3, K_4 lead to a situation similar to that shown in fig. 6.16: the line defining the recrystallized grain size lies entirely in the power-law-creep field, i.e. no change in creep mechanism is possible for realistic values of the starting grain size. However, there does not seem to be any clear physical reason why it should always be so in all cases (the recrystallized grain size is not in 'equilibrium' with stress). Schmid *et al.* (1977) found a situation where diffusion creep occurs in apparently recrystallized fine-grained limestone.

6.4   Geological palaeopiezometers

An important problem in geology and geophysics consists in finding an estimate for the applied stress that caused the visible deformation of crust and mantle rocks (mostly quartz and olivine). The existence of empirical relations between the applied stress and various microstructural parameters (dislocation density, subgrain size, recrystallized grain size) in the case of metals spurred the geological community to investigate whether these relations were also valid for minerals. Experiments on creep and dynamic recrystallization of minerals, as we have already seen, yielded similar relations. The next step was to use these empirical relations to estimate the applied stress from microstructural measurements on naturally deformed rocks. The so-called 'palaeopiezometers' were then used to estimate the palaeo-stress in crustal shear zones (e.g. Kohlstedt & Weathers, 1980; Christie & Ord, 1980) or constructing stress profiles in the lithosphere (Mercier *et al.*, 1977; Mercier 1980) from the microstructure of peridotite xenoliths. The three palaeopiezometers – dislocation density ρ proportional to σ^2, subgrain size proportional to σ^{-1}, recrystallized grain size proportional to $\sigma^{-1.2}$ – have been repeatedly assessed in the literature (Nicolas & Poirier, 1976; White, 1979; Ross *et al.*, 1980). We will only review here the principal problems that arise when the laboratory empirical relations are applied to naturally deformed rocks.

6.4.1 Dislocation density

Although the relation $\rho \propto \sigma^2$ is found to be valid in most experiments where the stress has been held at a constant value, its

Fig. 6.16. Recrystallized grain size and creep mechanisms. The boundaries between creep domains (PLC – power-law creep; NHC – Nabarro–Herring creep; CC – Coble creep) have been calculated for olivine using Ashby & Verrall (1978) rheological equations. The recrystallized-grain-size–stress line (dashed) is taken from Karato *et al.*'s (1980) experiments; it lies entirely in the PLC field for reasonable values of grain size and stress. An example of grain-size evolution from an initial state (circle) is given.

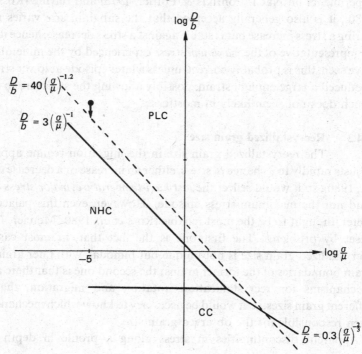

application to naturally deformed rocks poses many problems (White, 1979; Nicolas & Poirier, 1976), the principal of which has its source in the fact that the dislocation density is not very stable against changes in temperature and stress. As a result, the average dislocation density in a deformed monomineralic rock most often reflects a late stage of stress-free anneal or a low-temperature high-stress emplacement stage, but it is seldom representative of the main strain-producing episode, except when it is a major episode and when all palaeopiezometers give the same answer (as was apparently the case for the quartzite in the shear zone investigated by Kohlstedt & Weathers, 1980). In the majority of the cases, the stress found by using the dislocation density is unreliable and sometimes quite high (Christie & Ord, 1980).

6.4.2 Subgrain size

As we have seen, there are several scales of subgrains, and measuring the size of small subgrains by transmission electron microscopy yields a higher stress than measuring the size of the larger subgrains seen optically. There is now a tendency to consider that the larger, more misoriented subgrains visible by decoration or optically are the only ones that are representative of the major straining episode. From laboratory experiments on NaCl (Pontikis & Poirier, 1975) and olivine (Ross *et al.*, 1980), it is also generally accepted that the subgrain size varies rapidly during a stress increase but is stable against a stress decrease, hence it would be representative of the *maximum stress* experienced by the mineral. As we have seen, this is probably correct, unless a later episode at lower stress has produced a large enough strain, possibly allowing the structure to evolve, which does not seem likely in most cases.

6.4.3 Recrystallized grain size

The recrystallized grain size in the migration regime apparently adjusts rapidly to a change of stress, either an increase or a decrease (Ross *et al.*, 1980), so it would reflect the *last high-temperature, major stress episode* (and not the maximum-stress episode). However, even this palaeopiezometer, thought to be the most reliable (Ross *et al.*, 1980; Mercier, 1980), is beset by problems. The first one is the fact that, in most cases, the recrystallized grain size is not unique but bimodal, with finer grains near grain boundaries of the former grains; the second one is that there are two mechanisms for recrystallization, rotation and migration, that yield different grain sizes, so it would be necessary to know which mechanism has been responsible for the observed grain size.

Apparent discontinuities in stress along a profile in depth in the lithosphere might be spurious and result from the crossing at some depth of the boundary between two recrystallization regimes giving different grain sizes (Mercier, 1980).

Finally, in many cases (quartzites, marbles) the grain boundaries may be stabilized against migration by a dispersed second phase (generally flaky minerals like mica that lie along grain boundaries); the grain size, then, has no relation whatsoever with the applied stress (White, 1979).

7

Diffusion creep, grain-boundary sliding and superplasticity

At high temperature and low stresses, deformation of fine-grained materials proceeds by mutually accommodating grain-boundary sliding and transport of matter. According to which mechanism provides the greater amount of strain, one speaks of diffusion creep or grain-boundary sliding but they are never really dissociated.

Diffusion creep was predicted theoretically before being observed. If the transport of matter occurs by lattice diffusion, it is Nabarro–Herring creep and the viscosity varies as the grain size squared; if the transport of matter occurs by grain-boundary diffusion, it is Coble creep and the viscosity varies as the grain size raised to the third power.

Grain-boundary sliding accommodated by diffusion creep is described by the same kind of equations as diffusion creep. In most materials, if the grain size is small and stable there is a domain of strain-rate where the strain-rate sensitivity is higher than for dislocation creep and where tensile deformation can take place in a stable manner up to very large strains: it is the superplastic domain. Models for superplastic flow account for the high strain-rate sensitivity and the possibility of large strains, by grain-boundary sliding during grain-shifting events, locally accommodated by diffusion creep or climb and glide of grain-boundary dislocations in the mantle of the grains.

7.1   Generalities

Intragranular creep by dislocation motion is not the only high-temperature deformation mode of crystals. We have already mentioned (§ 2.1) that strain can be achieved by transport of matter by diffusion (diffusion creep) or by shear along the grain boundaries (grain-boundary sliding or GBS). Intragranular dislocation slip may not be uniform and we have seen that the plastic incompatibilities are usually relieved by geometrically necessary dislocation which can rearrange into polygonization walls, but intragranular creep in a polycrystal, on the whole, need not be otherwise accommodated. Such is not the case for diffusion creep or grain-boundary sliding. GBS creates voids or overlaps that have to be accommodated by diffusion creep, and diffusion creep, by deforming the grains, also creates voids or overlaps that must be accommodated by GBS (fig. 7.1). Indeed, *diffusion creep creates the driving force for GBS and vice versa. These processes are therefore strongly coupled and mutually accommodating; one cannot exist without the other.* However, it does not necessarily follow that both carry the same amount of strain (as in climb-controlled dislocation creep, for instance, climb provides little strain as compared with slip). We will therefore consider the two end-members of a continuum: GBS-controlled diffusion creep, in which most of the strain is due to transport of matter, and diffusion-creep-controlled GBS, in which most of the strain is due to GBS. We will then deal with the case of superplasticity, which also proceeds by mutually accommodating GBS and diffusion creep, but is usually studied separately on account of its low stress sensitivity of the strain-rate which imparts to superplastic materials the property of deforming up to large tensile strains without plastic instability.

7.2 Diffusion creep

7.2.1 Nabarro–Herring creep

Nabarro (1948) was the first to propose that a non-hydrostatic stress field could give rise to different vacancy concentrations on surfaces of a crystal with different orientations with respect to the stress field, hence cause a flux of vacancies between surfaces and a flux of ions in the opposite sense. Thus the transport of matter induced by the non-hydrostatic stress field would cause a pure shear deformation, making the applied stress do

Fig. 7.1. Relation between grain-boundary sliding and diffusion creep. In a model two-dimensional polycrystal deformed in pure shear, GBS creates voids that must be filled by diffusion creep or conversely diffusion creep causes GBS.

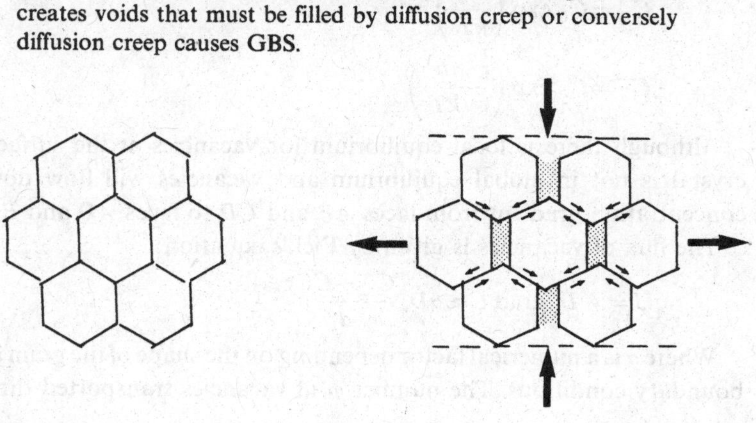

work on the sample. Obviously this process can be of importance only if the flux of vacancies is large, i.e. if there are enough sources (grain boundaries) and if the vacancy mobility is high enough. As a result, diffusion creep would be expected at very high temperatures and for very small grain sizes; the latter requirement also implies that dislocation sources cannot operate in small grains (unless the stress is quite high) and so diffusion creep would not be swamped by dislocation creep. Nabarro also pointed out that diffusion creep could well be effective in the Earth's mantle, an idea that was later taken up by Gordon (1965).

Diffusion creep stands alone among the creep processes, in that a rheological equation was first theoretically proposed and the phenomenon was actually proved to be real only 15 years later when Squires *et al.* (1963) and Harris & Jones (1963) noticed it in a Mg 0.5% Zr alloy (fig. 7.2). We will first give a simple calculation (following Nabarro) leading to the mechanical equation of state for diffusion creep; then we will briefly consider the more elaborate thermodynamic treatments and more realistic boundary conditions.

Let us consider the ideal case of a small isolated cubic crystal (of size d), in a pure shear-stress regime (fig. 7.3). Let us assume that vacancies are only created and absorbed at the surfaces and that they can diffuse in the bulk with a diffusion coefficient D_v.

At a given temperature T, the thermal equilibrium concentration of vacancies in the crystal is $C_0 = N_v/b^3$, where N_v is the equilibrium atomic fraction and $b^3 = \Omega$ is the atomic volume (see §2.2.1). Creation of a vacancy at the face AB (fig. 7.3) is helped by the tensile stress, whereas it is necessary to work against the compressive stress to create a vacancy at face BC. The equilibrium concentrations of vacancies at the faces AB and BC, respectively, are:

$$C^+ = C_0 \exp\left(\frac{\sigma b^3}{kT}\right) \tag{7.1}$$

$$C^- = C_0 \exp\left(-\frac{\sigma b^3}{kT}\right) \tag{7.2}$$

Although there is local equilibrium for vacancies at the surfaces, the crystal is not in global equilibrium and vacancies will flow down the concentration gradient from faces AB and CD to faces AD and BC.

The flux of vacancies is given by Fick's equation:

$$J = -D_v \text{ grad } C \approx \alpha D_v \frac{C^+ - C^-}{d} \tag{7.3}$$

Where α is a numerical factor depending on the shape of the grain and the boundary conditions. The number ϕ of vacancies transported through a

Fig. 7.2. Diffusion creep in Mg 0.5% Zr alloy. Annealing 1 h at 600 °C in H_2 created ZrH_2 precipitates. Material transported by diffusion to the grain boundaries normal to the tensile stress is free from precipitates and appears as denuded zones ($\dot{\varepsilon} = 7 \times 10^{-8}$/s, $T = 500$ °C, $\sigma = 350$ g/m^2) (scale bar = 0.1 mm). (Courtesy R. B. Jones.)

face of area d^2 per second is:

$$\phi = Jd^2 \tag{7.4}$$

To the flow of vacancies corresponds a flow of atoms in the opposite sense so that forming a vacancy corresponds to letting an atom (of volume b^3) pop out at the surface. If the total volume ϕb^3 arriving per second at a surface is spread evenly, it creates a layer of thickness $\phi b^3/d^2$ and the sample therefore elongates with a strain-rate:

$$\dot{\varepsilon} = \frac{1}{d} \cdot \frac{\phi b^3}{d^2} = \phi \frac{b^3}{d^3} \tag{7.5}$$

From (7.5), (7.4), (7.3), (7.2), (7.1) we may write:

$$\dot{\varepsilon} = \alpha \frac{D_v C_0 b^3}{d^2} \sinh\left(\frac{\sigma b^3}{kT}\right) \tag{7.6}$$

For small stresses, $\sigma b^3 \ll kT$, we can write:

$$\dot{\varepsilon} = \alpha \frac{D_v C_0 b^3}{d^2} \cdot \frac{\sigma b^3}{kT} \tag{7.7}$$

as $D_v C_0 b^3 = D_v N_v = D_{sd}$ coefficient of self-diffusion, we finally obtain:

$$\blacktriangleright \qquad \dot{\varepsilon} = \alpha \frac{D_{sd}\sigma\Omega}{d^2 kT} \tag{7.8}$$

We see that diffusion creep is Newtonian viscous ($\dot{\varepsilon} \propto \sigma$) and that the viscosity is grain-size-dependent, which is not the case for dislocation creep.

Herring (1950, 1951) rigorously formalized the problem of creep by

Fig. 7.3. Principle of Nabarro–Herring creep. The concentration of vacancies at faces in tension C^+ is higher than in bulk C_0, at faces in compression $C^- < C_0$. Vacancies flow from faces in tension to faces in compression and matter flows in the opposite sense.

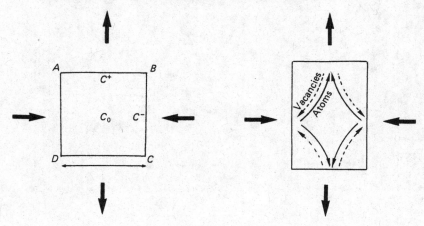

lattice diffusion in a polycrystal, which earned it the name: Nabarro–Herring creep.

Herring expressed the flux of matter in terms of the chemical potential (or partial molar free energy) of atoms and vacancies. The diffusion flux, in atoms per unit area and second is:

$$J = \frac{v}{\Omega}$$

where the velocity v is given by Einstein's equation:

$$v = \frac{DF}{kT} \tag{7.9}$$

The driving force F is:

$$F = -\text{grad}\,(\mu - \mu_v) \tag{7.10}$$

where μ and μ_v are respectively the chemical potential of atoms and vacancies. Hence:

$$J = -\frac{D}{\Omega kT}\,\text{grad}\,(\mu - \mu_v) \tag{7.11}$$

It is then reasonably assumed that the flux is conservative, i.e. div $J = 0$ inside the crystal; it follows that the Laplacian of $(\mu - \mu_v)$ is also equal to zero:

$$\nabla^2(\mu - \mu_v) = 0 \tag{7.12}$$

It is then possible to determine the diffusion flux, if the value of $(\mu - \mu_v)$ is known at the boundaries. Note that the values of $(\mu - \mu_v)$ determined inside the crystal are useful to find the diffusion flux but have no thermodynamical meaning, as the chemical potential of the atoms of a solid under non-hydrostatic stresses can be defined only at the surface (see Kamb, 1961; Paterson, 1973); the same is true of the chemical potential of the vacancies. At the surfaces, $(\mu - \mu_v)$ can be expressed as a linear function of the normal stress on the surface. Herring finally finds an expression for $\dot{\varepsilon}$ identical to eq. (7.8).

Herring considered an ideal polycrystal formed of identical spherical grains, for the cases where tangential stress relaxation is possible or not at the grain boundaries (i.e. possibility or impossibility of GBS); the coefficient α takes different values for the two cases:

$$\alpha = 16 \quad \text{if} \quad \text{GBS is impossible}$$
$$\alpha = 40 \quad \text{if} \quad \text{GBS is possible}$$

In both cases, the stress σ appearing in eq. (7.8) is the shear stress; if one uses the engineering stress in tension or compression, the coefficient must be replaced by $\alpha/3$.

N.B. It is easy to convince oneself of the necessity of the factor 1/3 by using the correspondence principle between elasticity and viscosity. For shear stresses, we have $\sigma = \mu\varepsilon$ for elastic solids and $\sigma = \eta\dot{\varepsilon}$ for viscous fluids, and, for engineering normal stresses, we have $\sigma = E\varepsilon$ and $\sigma = \eta'\dot{\varepsilon}$ respectively. Now for incompressible elastic solids ($\nu = 0.5$) we have $E = 3\mu$, hence for incompressible viscous fluids $\eta' = 3\eta$.

7.2.2   Coble creep

Coble (1963) investigated the case of a polycrystal where transport of matter would occur not by lattice diffusion but by grain-boundary diffusion. Assuming that the grains were spherical (as did Herring) and that GBS was possible he found:

▶
$$\dot{\varepsilon} = \frac{148 D_{gb} \delta\sigma\Omega}{\pi d^3 kT} \tag{7.13}$$

where D_{gb} is the grain-boundary diffusion and δ the grain-boundary thickness. Formula (7.13) applies to the case where σ is the engineering stress; for σ being a shear stress we would replace σ by 3σ, hence:

$$\dot{\varepsilon} = \frac{141 D_{gb} \delta\sigma\Omega}{d^3 kT} \tag{7.14}$$

The important difference between Nabarro–Herring and Coble creep lies in the fact that, for the latter, the creep-rate depends on the inverse of d^3 rather than the inverse of d^2 and also that it depends on the grain-boundary diffusion. The activation energy for grain-boundary diffusion being smaller than for lattice diffusion, Coble creep can be important at temperatures less high than those necessary for Nabarro–Herring creep (fig. 7.4). The geologically important phenomenon of deformation by pressure solution, with transport of matter in a liquid film at the grain boundaries, can be treated exactly as Coble creep (McClay, 1977).

7.2.3   Experimental observations

As already mentioned, Nabarro–Herring creep was first observed in Mg 0.5% Zr alloys (Squires *et al.*, 1963); Harris & Jones (1963) calculated the strain from the width of the denuded zones near the grain boundaries in tension and found a good agreement with Herring's formula.

Nabarro–Herring creep was also demonstrated in 'zero creep' experiments on thin wires (see Jones, 1969). Later, Nabarro–Herring and Coble creep were found to be active in many metallic and ceramic systems (see Burton, 1977, for an extensive review).

In most cases there is a transition stress for a given grain size, below which the Newtonian creep takes over from dislocation creep as can be expected (figs. 7.5 and 7.6). There obviously is also a critical grain size below which power-law creep gives way to diffusion creep. Raj & Ghosh (1981) have shown that a more realistic bimodal distribution of grain sizes in a material gives rise to a transition regime spread out over three or four orders of magnitude in stress-rate. Experiments at very low stresses show that there often exists a small threshold stress below which no creep occurs, i.e. that the viscous flow is of Bingham rather than Newton type (Crossland *et al.*, 1973; Crossland, 1974). The fact that the grain boundaries are not perfect sources or sinks for vacancies, i.e. that there can be some degree of interface-reaction control of the creep-rate, may be at the origin of the threshold stress (Ashby, 1969).

Subgrain boundaries may also act as sources and sinks of point defects, as originally suggested by Friedel (1964), and Coble creep may occur by transport of matter along the subgrain boundaries by pipe diffusion; indeed, in some cases there is a better agreement between the observed and the calculated viscosity if the latter takes into account the subgrain size rather than the grain size (Crossland, 1974); the observed viscosity, however, remains smaller than the calculated one, even after this modification.

Fig. 7.4. Nabarro–Herring and Coble creep fields in the strain-rate–temperature–grain-size space.

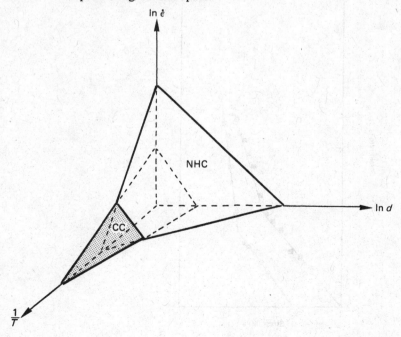

Fig. 7.5. Linear $\dot{\varepsilon}$–σ plot for dislocation creep and diffusion creep. Diffusion creep dominates at stresses lower than σ_t (for constant grain size).

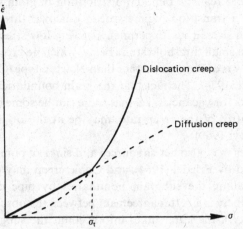

Fig. 7.6. Log $\dot{\varepsilon}$–log σ plot for creep of β-cobalt. At low σ, diffusion creep ($n=1$); at high σ, power-law creep ($n>1$). From top to bottom, temperatures and grain sizes are 760°, 105 μm, and 620°, 46 μm (after Sritharan & Jones, 1979).

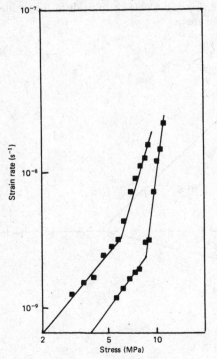

7.3 Mutually accommodating grain-boundary sliding and diffusion creep

Herring (1950) and Coble (1963) calculated the diffusion creep-rate of an isolated spherical grain with extreme boundary conditions on the tangential stress at the surface (relaxation or absence of relaxation). The extension to the real polycrystals of the mechanical equations of state is not straightforward and must take into account the fact that the grains are polyhedra that must remain in contact during compatible plastic deformation. Lifshitz (1963) first analysed the conditions for compatibility at the grain boundaries, taking into account the necessity for grain-boundary sliding as well as the consequences of the existence of shorter diffusion paths near the edges of the grains. The coupling between diffusion creep and GBS was analysed by Raj & Ashby (1971), Stevens (1972) and Gifkins (1976). Raj & Ashby considered the general problem of sliding at a non-planar grain boundary; they considered in particular the cases of purely elastic and of diffusional accommodation of GBS. They modelled the shape of the boundaries of two-dimensional grains by a Fourier series; fig. 7.7 shows the elementary sinusoidal shape of a boundary subjected to a shear stress and accommodating the sliding by local diffusional creep. They showed that GBS in a model polycrystal of hexagonal tiles with non-planar boundaries can be accommodated elastically only over very short distances (≈ 50 Å) and that GBS may proceed only if there is local accommodation by plastic slip of dislocations or by transport of matter.

Considering transport of matter both by lattice and grain-boundary diffusion, Raj & Ashby find a mechanical equation of state for diffusion-creep-controlled GBS of a two-dimensional polycrystal of grain size d, with sinusoidal grain boundaries:

$$\blacktriangleright \qquad \dot{\varepsilon} = \alpha \frac{D_{\text{eff}} \sigma \Omega}{d^2 kT} \qquad (7.15)$$

Fig. 7.7. Grain-boundary sliding accommodated by diffusion creep for a sinusoidal grain boundary (after Raj & Ashby, 1971).

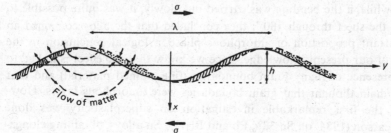

The coefficient α has the same value as in eq. (7.8) for Nabarro–Herring creep when tangential stresses are relaxed at the boundary: $\alpha = 40$. D_{eff} is an effective diffusion coefficient:

$$D_{eff} = D_{sd}\left[1 + \frac{\pi\delta}{\lambda}\frac{D_{gb}}{D_{sd}}\right] \tag{7.16}$$

where D_{sd} and D_{gb} are the coefficients of lattice and grain-boundary self-diffusion respectively, δ is the grain-boundary thickness and λ the wave length of the sinusoidal profile of the grain boundaries (or basic periodicity of the boundaries). Note that for lattice diffusion alone, (7.16) gives the same expression for creep-rate as the Herring theory and for grain-boundary diffusion alone (7.16) is practically identical with the Coble expression (7.14) if $\lambda \approx 2d$.

7.4   Structural superplasticity

7.4.1   Generalities

Some fine-grained materials, subjected to tensile tests at high temperature and low strain-rate, may stretch without necking, up to unusually high elongations of the order of 1000%; they are then said to behave superplastically. Superplasticity is a behaviour, not a definite phenomenon like superfluidity in liquid helium; its definition is therefore loose, and there is no unique theory of superplasticity.

The possibility of reaching very large elongations without necking, i.e. of deforming in a stable manner, is a manifestation of a high strain-rate sensitivity of the stress (high m, see § 1.3). Glasses, pitch, and 'silly putty' are typical examples of solids (generally amorphous) that behave in a brittle fashion under a shock (the high stresses reached at high strain-rates relax by fracture) and flow viscously without instabilities if the load is applied slowly.

Rosenhain *et al.* (1920) were the first to discover that some metallic alloys could behave in the same way as pitch: if the ternary eutectic alloy Zn 4% Cu 7% Al was 'bent rapidly or given a blow, it broke without apparent flow while, if the bending was carried out slowly, it was quite possible to bend the sheet through 180°'; they concluded that the alloy contained an important proportion of amorphous phase, a logical conclusion in the light of our present knowledge, since we know that the effect is linked to the presence of many grain boundaries (fine-grained material) and that Rosenhain thought that grain boundaries were amorphous layers. However, the first remarkable investigation on superplasticity was done by Pearson (1934) on Sn 38% Pb and Bi 44% Sn alloys, obtaining elongations of nearly 2000%. He clearly saw most of the characteristics of superplasticity, and recognized the importance of grain size.

Most of the superplastic alloys were two-phased eutectics and the high-temperature superplasticity was later attributed to the phase transformation, rather than to the resulting microstructure, by the Soviet investigators who were active in the field of transformation plasticity (see chapter 8). The confusion was spread in the West by the review of Underwood (1962); it was dispelled by the work of Backofen and his team at MIT who investigated numerous eutectic and eutectoid systems: e.g. Al 22% Zn (Backofen *et al.*, 1964), Sn 38% Pb (Avery & Backofen, 1965).

Superplasticity was later found in many metallic systems, e.g. Ni—Fe—Cr alloys (Hayden *et al.*, 1967) (see recent reviews by Edington *et al.*, 1976; Taplin *et al.*, 1979; Padmanabhan & Davies, 1980).

It is, however, not enough for a material to possess a very low value of strain-rate sensitivity to be effectively superplastic. If such were the case, Newtonian diffusion creep, with $m = 1$, which is a stable deformation mode, would always lead to superplasticity; it obviously does not, for the only reason that the creep-rate is very low (it would take a year to obtain an elongation of 1000% at the creep-rate of $3 \times 10^{-7}/s$ typical of Nabarro–Herring creep). Even before looking into the possible mechanisms for superplasticity, we can predict that they will belong to the class of grain-boundary sliding, possibly accommodated by diffusion creep, for only GBS can provide the large strains needed, while local accommodation by diffusion creep involves a shorter diffusion path than when diffusion creep is the main strain-producing mechanism.

In the Earth Sciences community, one never deals with large elongation of minerals and rocks but rather with deformation in simple shear or compression; the criterion of stable deformation without necking is therefore almost irrelevant; nevertheless, the word 'superplasticity' has unfortunately been introduced to refer to diffusion-creep (or otherwise) accommodated grain-boundary sliding, which indeed is at the origin of superplasticity, when there is superplasticity. In that sense, 'superplastic flow' has been experimentally demonstrated to occur in Carrara marble deformed in compression (Schmid *et al.*, 1977) and it has been inferred, on a microstructural basis, in some mylonites (Boullier & Gueguen, 1975).

7.4.2 Characteristics of the superplastic regime

Any model of superplasticity or 'superplastic flow' must account, at least partially, for the following features:

(i) The strain-rate sensitivity m is higher than for dislocation creep in the superplastic domain: $0.5 \leqslant m \leqslant 0.7$ instead of $m \approx 0.2$ for dislocation creep.

Over the entire experimental range of strain-rates, the curve $\log \sigma$ against $\log \dot{\varepsilon}$ is usually sigmoidal (fig. 7.8), the superplastic domain corresponds to the part of the curve near the inflexion point where $m = \mathrm{d} \log \sigma / \mathrm{d} \log \dot{\varepsilon}$ is maximum, at lower strain-rates than the dislocation-creep domain. As the value of m is an important clue to the existence of superplasticity, experimental methods for determining it must be critically assessed (Nuttall, 1971). There are two methods for determining m; by strain-rate jump during a tensile test, or by plotting $\log \sigma$ against $\log \dot{\varepsilon}$ for various experiments at the same strain and taking the slope of curve. The two methods usually do not yield the same value for m, except when strain-hardening is zero ($\gamma = 0$, see §1.2.3) (Tonejc & Poirier, 1977).

(ii) The $\log \sigma$–$\log \dot{\varepsilon}$ curve, hence the domain of superplasticity, is displaced towards higher strain-rates as the grain size decreases (fig. 7.9). A small grain size is therefore essential to obtain superplasticity for reasonable strain-rates.

(iii) The grains usually remain equiaxed, even after very large strains, i.e. the elongation of the grains, if any, is much smaller than the total elongation of the sample; there is usually no dislocation substructure inside the grains and the latter preferred orientation is weak or absent. These observations point to grain-boundary sliding as the main source of strain in the

Fig. 7.8. Superplastic regime. Log σ–log $\dot{\varepsilon}$ curve is sigmoidal (for a given grain size). The high strain-rate sensitivity m in the region of the inflexion point corresponds to superplasticity.

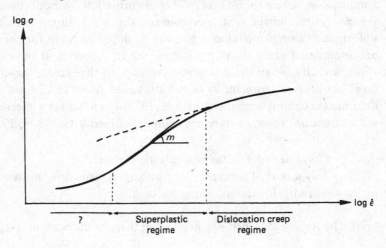

superplastic regime. Grain-boundary sliding is indeed important and often accompanied by rotation of the grains (Taplin *et al.*, 1979): grains in the process of switching neighbours have also been observed *in situ* in ZnAl specimens deformed in a 1 MeV electron microscope (Naziri *et al.*, 1973, 1975). Superplasticity has been obtained at room temperature in fine-grained aluminium wetted by liquid gallium (Marya & Wyon, 1975): the liquid gallium forms a low-melting-point eutectic layer at the grain boundaries and allows easy GBS.

Most superplastic alloys are fine-grained two-phase eutectics with a comparable volume fraction of each phase (*microduplex texture*); the role of interfaces between phases must therefore be considered. It seems that the essential function of the microduplex texture is to prevent the grain growth of both phases. The topology and the connectivity of the phases are obviously important factors, unfortunately little investigated; it must be pointed out that they can only be known by a tedious method of serial section of the samples (and not by simple inspection of a few micrographs): the CdZn eutectic was investigated in this manner (Tonejc & Poirier, 1977); it was found that the harder Zn-rich phase was unconnected despite its high volume fraction (26%) and was present as spheroidal globules at fourfold junctions, probably due to interfacial energy effects; the role of the interfaces between phases was probably very small in this case.

It must also be noted that the presence of a second phase is not necessary to superplasticity: for example, Gifkins (1976) found that solid-solution

Fig. 7.9. Superplasticity in Cd 25.6% Zn eutectic alloy ($T = 210°$C): log σ–log $\dot{\varepsilon}$ curves for various grain sizes can be brought together by a translation in the direction of line t. At constant stress, $\dot{\varepsilon} \propto d^{-2.2}$. The grain sizes in μm are 3, 7, 10 for curves 1, 2, 3 respectively (after Tonejc & Poirier, 1977).

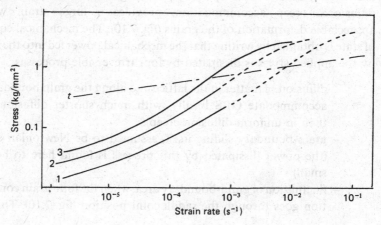

alloys of lead and thallium (mainly Pb 2.5% Tl) exhibited superplasticity; the stability of the grain size was attributed to the confinement of activity to the 'mantle' leaving the 'core' of the grains undeformed and inhibiting recrystallization. Finally the activation energy of the strain-rate is often smaller than the lattice self-diffusion activation energy (see Mohamed & Langdon, 1976, for a discussion of the experimental methods for obtaining Q).

7.4.3   Microscopic models

Many microscopic models of superplastic behaviour have been proposed, which generally find some support in a somewhat wide spectrum of experimental results on various materials, while never quite succeeding in accounting for all observations. The major models have recently been reviewed and discussed by Gifkins & Langdon (1978). We will focus here only on those we find more interesting and fruitful in their approach. Among these, the model of Ashby & Verrall (1973) holds a special place, for, even though it has (correctly) been criticized over points of greater or lesser importance, it has introduced a new approach, which later authors have used and modified in their models.

Ashby & Verrall realized that it was impossible to account for the fact that grains remained equiaxed after very large strains by uniform deformation, since the elongation of all grains should be equal to the total elongation. This is obviously why we have to have grain-boundary sliding, but here it was introduced in a new form: the basic unit of deformation is thought to be a *neighbour-switching event*, as happens in the deformation of an oil emulsion consisting of 'grains' of oil separated by very thin boundaries of detergent. The change in configuration of a four-grain unit in a two-dimensional model polycrystal, by grain-boundary sliding with diffusional creep accommodation, provides a large strain without appreciable deformation of the grains (fig. 7.10). The mechanical equation of state is obtained by writing that the mechanical power fed into the system by the applied stress is dissipated by four irreversible processes:

- diffusion of matter in the lattice and along the grain boundaries to accommodate GBS locally (with much shorter diffusion paths than in uniform diffusion creep).
- grain-boundary sliding itself, assumed to be Newtonian viscous (the power dissipated by this process is found here to be quite small).
- fluctuation of grain-boundary area, when the four-grain configuration goes through the saddlepoint position (fig. 7.10). This term

introduces a small threshold stress, below which deformation is not possible. Although the existence of a threshold stress has been controverted, it still seems the easiest way to account for the decrease in m at very low strain-rates (sigmoidal σ, $\dot{\varepsilon}$ curve). Burton (1971) showed that, owing to the properties of the logarithmic plot ($\dot{\varepsilon} = 0$ is removed at infinity), a threshold stress must result in a sigmoidal curve (fig. 7.11).

– interface reactions, expressing the fact that grain boundaries are not perfect sources and sinks for vacancies.

In the simpler case where the condition for interface reaction is relaxed (perfect sources and sinks) the final equation takes the form:

$$\blacktriangleright \qquad \dot{\varepsilon} = 100 \frac{\Omega}{kTd^2} \left(\sigma - \frac{0.72\Gamma}{d^2} \right) D_{sd} \left(1 + \frac{3.3\delta D_{gb}}{dD_{sd}} \right) \qquad (7.17)$$

where Γ is the specific grain-boundary energy and all the other symbols have their usual meaning. With the exception of the threshold stress, eq. (7.17) looks very much like eq. (7.15) for GBS accommodated both by lattice and grain-boundary diffusion.

The model gives a Newtonian viscosity ($m = 1$), which is indeed seldom found in the superplastic domain: m is usually closer to 0.5 ($\dot{\varepsilon} \propto \sigma^2$). However, if one takes the interface reaction into account, the constitutive equation has an asymptotic form in which $\dot{\varepsilon} \propto \sigma^2$. This model has successfully explained a number of experimental results (e.g. the low-stress creep of fine-grained UO_2 (Chung & Davies, 1979)).

The major limitation of Ashby & Verrall's model is that it is two-dimensional; the neighbour-switching event that is at the core of the model obviously can take a different topological configuration in a real three-dimensional crystal. Besides, the model conserves the area in two dimen-

Fig. 7.10. Neighbour-switching event in the superplastic regime: a group of four grains suffers a 55% strain without deformation of the grains, except locally in the intermediate state to accommodate grain-boundary sliding (after Ashby & Verrall, 1973).

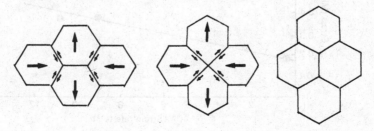

sions (equivalent to conserving the volume in three dimensions), but this property is not easily got rid of in extending the model to three dimensions where the area is not conserved.

Gifkins (1978) proposed another grain-switching model which takes into account the third dimension and made it the central element of a new model that uses basically the same approach as a previous one (Gifkins, 1976). The main points are the following.

Strain is always produced by GBS, but here, rather than assuming a

Fig. 7.11. The sigmoidal shape of the log σ–log $\dot{\varepsilon}$ curve may be spurious and due to the removal of $\dot{\varepsilon} = 0$ to infinity in the logarithmic scale (*a*). Plotting on a linear scale makes a threshold stress apparent (*b*). Data for Pb–Sn eutectic at RT (after Burton, 1971).

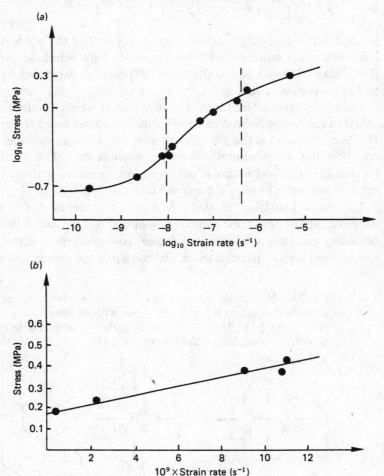

Newtonian viscosity, Gifkins considers the motion of grain-boundary dislocations as the agent of GBS. The sliding produces stress concentration at the triple edges, which gives rise to a σ^2 dependence of the strain-rate. Accommodation of GBS in the superplastic domain takes place in the 'mantle' of the grains by grain-boundary dislocations dissociating into crystal dislocations and then gliding and climbing. This flow in the mantle of width $\approx 0.07d$ allows grains to rotate. At the same time, grains move in the third dimension so that a grain may appear from above or below at the centre of a four-grain unit (fig. 7.12), as observed by Naziri *et al.* (1973, 1975).

Fig. 7.12. Elongation by grain-boundary sliding and accommodation by emergence of grains from one layer to the next. A gap forms between four grains and is filled by an emerging grain (after Gifkins, 1978).

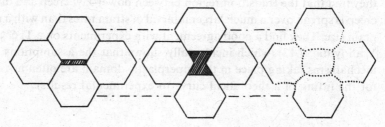

Fig. 7.13. Effect of a mixture of grain sizes on superplastic flow: log σ–log $\dot{\varepsilon}$ curves calculated for two grain-size distributions (insert). Circles correspond to the experimental data for Ti 6% Al 4% V alloy (after Ghosh & Raj, 1981), $T = 1200$ K.

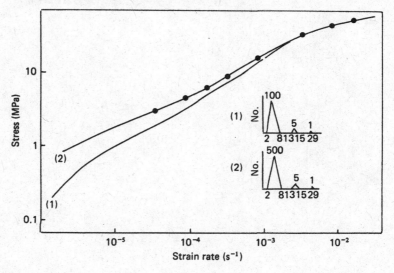

The resulting equation has the form:

▶
$$\dot{\varepsilon} = \frac{64\Omega D_{gb}\sigma^2}{\mu k T d^2} \qquad (7.18)$$

All models usually consider a unique grain size d, an obviously unrealistic assumption. Ghosh & Raj (1981) investigated the consequences of the existence of a distribution of grain sizes. However, they started from simple assumptions on the flow mechanism, ignoring grain-boundary sliding and using for the $\sigma–\dot{\varepsilon}$ curve an equation that simply combines the contributions of Coble creep and power-law creep. Also, they assumed that the average strain-rate of each grain was the same as the imposed strain-rate. Despite the inappropriateness of these assumptions to the superplastic domain, the model has some merit in showing the influence of a grain-size distribution. Assuming a mixture of grain sizes with a predominance of larger grains, they find that the transition region between power-law creep and diffusion creep is spread over a much larger interval of strain rates than with a unique grain size. They find a good agreement with experiments on a Ti 6% Al 4% V alloy (fig. 7.13) (which incidentally shows that the assumptions on the mechanisms taking place in the superplastic domain are often not critical for the fitting of a theoretical curve to experimental results).

8

Transformation plasticity

Transformation plasticity is a weakening of the mechanical properties of a polycrystal while it is undergoing a phase transformation. It must be distinguished from other processes linked to phase transitions, like the so-called 'Transformation-Induced Plasticity' which makes steels tougher.

After a brief review of the thermodynamics and kinetics of phase transformations, the experimental evidence for transformation plasticity in metals and ceramics is presented. The macroscopic model of Greenwood & Johnson, inspired from 'yielding creep' models, is reviewed: it considers that the internal stresses caused by the volume change of grains overcome the yield strength of the solid and make it flow under small stresses. Microscopic models are presented, in which the internal stress is relaxed by dislocations that move under the applied stress.

8.1   Introduction

We will define transformation plasticity as a weakening of the mechanical properties of polycrystal, *while* it is undergoing a phase transformation. The manifestations of transformation plasticity are:

  (i) an enhanced creep-rate, above the thermal creep-rate, in the case of creep at constant stress.
 (ii) a stress drop, in the case of constant strain-rate tests.

These manifestations cease when the phase transformation is over. Transformation plasticity, therefore, has nothing to do with the change in mechanical properties *consecutive* to a phase transition (e.g. the creep-rate of α-Fe changes to the creep-rate of γ-Fe after the α–γ transformation is completed); it is really a manifestation of the interaction between phase transformations and plastic deformation.

The definition we have given also excludes an important but very different phenomenon: Transformation-Induced Plasticity (TRIP), which is at the basis of the interesting properties of the TRIP steels (Zackay *et al.*,

1967). Low-alloy steels (generally close to a composition Fe 9 wt.% Cr 8% Ni 4% Mo 2% Mn 2% Si 0.25% C) are usually quenched, tempered and heat-treated to achieve higher tensile strengths, but the improvement in tensile strength is accompanied by a deterioration of the ductility, i.e. necking occurs at low strains causing early failure in tension by plastic instability. The instability appears because the strain-hardening coefficient is not high enough (see § 1.3). The idea of Zackay & Parker was to introduce strong barriers (stronger than dislocation tangles) to dislocation motion. The quenched and tempered austenitic alloy are thermodynamically metastable and straining favours the formation of martensite; strain-induced martensite plates act as the desired strong barriers, the strain-hardening coefficient goes up and elongations at rupture can reach 25%. The treatment conducive to martensite formation must be applied *during* straining, as it is the strain-hardening coefficient that must be raised.

It is therefore clear that the increase in ductility at no point involves a lowering of the flow stress; on the contrary, the idea of TRIP is to have at the same time a high strength and a high strain-hardening.

The transformation plasticity that constitutes the subject of the present chapter is sometimes also called by the unwieldy name, 'transformational superplasticity'; it was first described by Sauveur (1924) in iron bars submitted to torsion tests in a temperature gradient: twisting was concentrated in the portions of the bar where the α–γ transition occurred; Sauveur clearly stated that it seemed likely to him that 'when iron undergoes its alpha–gamma transformation, it acquires a temporary plasticity, which greatly exceeds the plasticity of gamma iron at considerably higher temperatures'. Wassermann (1937) noticed transformation plasticity in Fe 30% Ni austenitic steels during martensite formation. The Soviet workers after Bochvar (see Presnyakov & Chervyakova, 1959) investigated Al–Zn and Al–Cu alloys during the entectoid precipitation; they coined the work 'superplasticity' for the effect they noticed and which they explained by a 'vigorous transfer of atoms by diffusion accompanying the process of precipitation'. As the phase transition also produces a fine-grained texture leading to structural superplasticity, it is not clear that they observed a real transformation-plasticity effect, although there most likely is one.

Greenwood & Johnson (1965) proposed a mechanical explanation for transformation plasticity, which accounted for the experimental results, but did not propose any microscopic mechanism.

The interest in transformation plasticity was never very active in the metallurgical community, but the Earth Sciences community has recently realized that the conjunction of phase transformations and high-temperature deformation in the tectonically active regions of the crust and in the mantle could make transformation plasticity a relevant mechanism (e.g. Gordon, 1971; Sammis & Dein, 1974; Parmentier, 1981; Poirier, 1982).

In the next section we will provide a short review of the concepts of phase transformations and kinetics, useful for our purpose; we will then proceed to review the experimental results and the models.

8.2 A short summary of thermodynamics and kinetics of phase transformations

8.2.1 Equilibrium thermodynamics

We will limit ourselves in this summary to polymorphic (or allotropic) phase transformations, i.e. those that involve only one component, which (in the simplest case) can change from one crystalline structure to another. In the pressure–temperature plane one can define the domains of stability of each phase, in which the free energy G is minimal for the stable phase.

The domains of each phase are separated by an equilibrium boundary, locus of the points in the P, T plane for which the free energies of both phases are equal. It must be kept in mind that the strain energy stored in a crystal as dislocations has to be taken into account in the calculation of the free energy as it will alter the equilibrium curve: thus a phase α more stable than β in given P, T conditions in the undeformed state can become less stable (higher G) if it is deformed. There are many, altogether arbitrary, schemes of classification of transformations (Roy, 1973; Rao & Rao, 1978); we will mention here only the concept of order of a transition due to Ehrenfest. For first-order transitions, there is a discontinuity in the first derivatives of the free energy – the entropy S and the specific volume V:

$$\left.\frac{\partial G}{\partial T}\right)_P = -S$$

$$\left.\frac{\partial G}{\partial P}\right)_T = V$$

For second-order transitions, only the second derivatives of G present a discontinuity. As we will see below, the transitions involved in transformation plasticity and, indeed, most of the important transitions, are first-order and are therefore characterized by a volume change ΔV and a latent heat $L = T\Delta S$.

At equilibrium $\Delta G = 0$, it follows that the slope of the boundary in the P, T planes is given by the *Clausius–Clapeyron* rule:

$$\frac{\mathrm{d}P}{\mathrm{d}T} = \frac{\Delta S}{\Delta V}$$

First-order transitions can be effected by nucleation and growth of the product phase, controlled by diffusion or by propagation of the interface by motion of dislocations. The latter mechanism is particularly relevant to phase transformations involving a change in the stacking of ions, from hexagonal close-packed (ABABA) to cubic close-packed (ABCABC) and vice versa; it has been observed for the ε–β cobalt transition or the würtzite–sphalerite (ZnS) transition and it is perhaps active in many close-packed systems, given suitable P, T conditions (e.g. olivine–spinel; Poirier, 1981).

Martensitic transformations are shear, diffusionless transformations, characterized by crystallographic orientation relations between parent and daughter phase and by the rapidity of the transformation. They usually are athermal and shear-stress- or strain-sensitive.

8.2.2 Kinetics

Thermally activated transformations proceed at a certain reaction rate when the system is maintained at constant P and T. The reaction rate $\mathrm{d}X/\mathrm{d}t$ is defined as the derivative with respect to time of the volume fraction of transformed phase $X = V_\beta/V_{0\alpha}$, where V_β is the volume of the transformed phase at time t and $V_{0\alpha}$ the initial volume of the parent phase.

The kinetics of the isothermal reaction is known when one knows $X(t)$, the expression of the volume fraction as a function of time; it is given as a solution of a differential equation which can take two essential forms:

$$\frac{\mathrm{d}X}{\mathrm{d}t} = K(T)[1 - X(t)] \tag{8.1}$$

The right-hand side of the equation is linear in X; the reaction is said to be first-order (nothing to do with the order of the transformation). If

$$\frac{\mathrm{d}X}{\mathrm{d}t} = K(T)X(t)[1 - X(t)] \tag{8.2}$$

the reaction is said to be second-order. (The right-hand side is of the second degree in X.) $K(T)$ is the reaction constant.

For a first-order reaction (corresponding to the exhaustion of nucleation sites or to unidimensional growth of one phase as in the thickening of plates), integration of (8.1) yields:

$$X(t) = 1 - \exp(-Kt) \tag{8.3}$$

The reaction-rate constant $K(T)$ has the dimension of t^{-1}. It is useful to define a characteristic time for the reaction $\tau = 1/K(T)$ (fig. 8.1).

In many cases, however, the reaction is more rapid than first-order and can be empirically represented by:

$$\frac{dX}{dt} = K(T)[(1-X)t^m] \tag{8.4}$$

and

$$X(t) = 1 - \exp\left[-K(T)t^n\right] \tag{8.5}$$

with $n = m + 1$. For the first-order reaction we had $m = 0$ and $n = 1$. It must emphatically be said that n in (8.5) *is not the order of the reaction, even though $n = 1$ for a first-order reaction.*

The kinetic equation (8.5) is said to be of the *Avrami–Johnson–Mehl type* (fig. 8.2). Different values of n correspond to different conditions of nucleation and growth.

Fig. 8.1. First-order reaction kinetics. Volume fraction of transformed phase plotted against time. τ is the characteristic time. $X = 1 - \exp(-t/\tau)$.

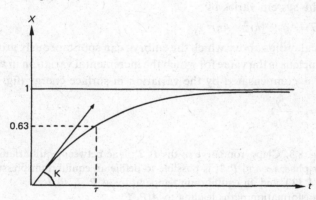

Fig. 8.2. Avrami kinetics. $X = 1 - \exp(-t/\tau)^n$. For small t (dilute transformed phase), $X \propto t^4$ (dashed curve).

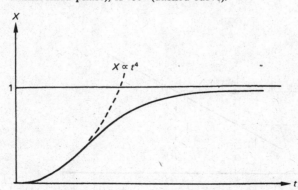

We will only consider here the elementary principles of the calculation of the volume fraction of transformed phase by nucleation and growth, when the transformed phase is still dilute (i.e. at the beginning of the transformation).

Let α and β be the transforming and transformed phase respectively; for α to transform to β, the P, T conditions must plot in the stability domain of phase, i.e. the free-energy variation between β and α: $\Delta g_{\beta\alpha}$ (per unit volume), must be negative. $\Delta g_{\beta\alpha}$ can easily be calculated, knowing P, T and the values of ΔV and ΔS for the transformation:

$$\Delta g_{\beta\alpha} = \Delta V(P - P_e) = \Delta S(T_e - T) \tag{8.6}$$

where P_e and T_e are the equilibrium pressure and temperature for constant T and constant P respectively (fig. 8.3). The formation of an embryo of phase β in α is accompanied by a decrease of free energy owing to the formation of a volume of β but also by an increase due to the formation of an α–β interface with specific energy γ. For a spherical embryo of radius r, the total free energy of the system varies by:

$$\Delta G = \Delta g_{\beta\alpha} \cdot \tfrac{4}{3}\pi r^3 + 4\pi r^2 \gamma \tag{8.7}$$

The critical radius above which the embryo can spontaneously grow and become a nucleus is the value for which the incremental variation in volume free energy is compensated by the variation in surface energy (fig. 8.4)

$$r_c = -\frac{2\gamma}{\Delta g_{\beta\alpha}} \tag{8.8}$$

Fig. 8.3. Clapeyron curve in the P, T plane between equilibrium fields for phases α and P. It is possible to define an equilibrium pressure at T, $P_e(T)$, and an equilibrium temperature at P, $T_e(P)$, for transformation paths leading to $A(P, T)$.

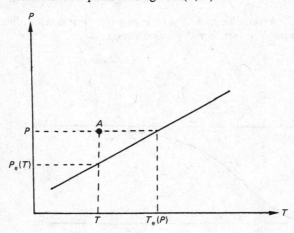

(For heterogeneous nucleation, on a dislocation or grain boundary for instance, eq. (8.7) should include an extra negative term representing the energy of the portion of defect suppressed by the presence of the nucleus.)

The energy barrier for nucleation is the value of ΔG for $r = r_c$

$$\Delta g_n^* = \frac{16}{3} \pi \frac{\gamma^3}{\Delta g_{\beta\alpha}^2} \tag{8.9}$$

and the nucleation rate (number of nuclei formed by unit time and unit volume) for homogeneous nucleation is:

$$I = N_0 v_n \exp\left(-\frac{\Delta g_n^*}{kT}\right) \tag{8.10}$$

As $\Delta g_n^* \propto 1/\Delta g_{\beta\alpha}$, the nucleation rate is higher far from the Clapeyron than close to it. The growth of the nucleus by $\mathrm{d}r$ causes the free energy of the system to decrease by:

$$\Delta g^* = \frac{\mathrm{d}}{\mathrm{d}r} \Delta G \cdot \mathrm{d}r \tag{8.11}$$

Fig. 8.4. Free energy against radius for a growing spherical domain of transforming phase. Below the critical radius r_c the embryo is not stable. Above r_c the embryo becomes a nucleus that can spontaneously grow, lowering ΔG for the system. Δg_n^* is the nucleation free-energy barrier.

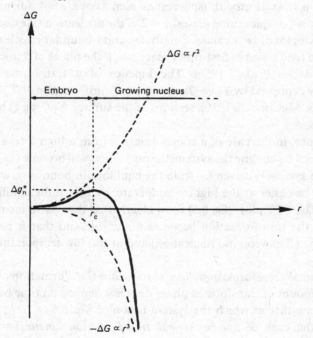

If there is an interface energy barrier to growth Δg_g^*, the growth rate by thermally activated fluctuations is:

$$\dot{R} = \frac{\mathrm{d}r}{\mathrm{d}t} \propto \exp\left(-\frac{\Delta g_g^*}{kT}\right)\left[1 - \exp\left(-\frac{\Delta g^*}{kT}\right)\right]$$

or, if $\Delta g^*/kT \ll 1$

$$\dot{R} \propto \frac{\Delta g^*}{kT} \exp\left(-\frac{\Delta g_g^*}{kT}\right) \tag{8.12}$$

For uniform nucleation, the number of nuclei formed during time $\mathrm{d}\tau$ is proportional to the untransformed volume V_x, to the nucleation rate I and to $\mathrm{d}\tau$; the transformed volume at time t resulting from the growth of all nuclei formed at times $\tau < t$ is therefore:

$$V_\beta = \int_{\tau=0}^{\tau=t} \tfrac{4}{3}\pi[\dot{R}(t-\tau)]^3 I V_x \, \mathrm{d}\tau \tag{8.13}$$

For a dilute system $V_\beta \ll V_x$, hence $V_x \approx V_{0x}$ and for $I = \mathrm{const.}$:

$$X = \frac{V_\beta}{V_{0x}} = \frac{\pi}{3} I \dot{R}^3 t^4 \tag{8.14}$$

The non-dilute case is more complicated, owing to the impingement of growing nuclei; in the general case, the Avrami equation can be theoretically justified and the exponent n depends on the geometry of growth; for a triaxial growth (spheres) as seen above, $n = 4$, for uniaxial growth (needles, or thickening plates), $n = 2$. If the nucleation rate decreases with time, n can take lower values. Finally, for grain-boundary nucleation it is possible to tend towards first-order reactions, if the plates of β thicken at grain boundaries (Cahn, 1956). The kinetics of a transformation is conveniently expressed by *Time–Temperature–Transformation (TTT) plots* (see Putnis & McConnell, 1980), sections of the surface $X(T, \log t)$ by $X = \mathrm{const}$ planes.

For instance, in the case of a transformation from a high T to a low T phase obtained by cooling the system, the nucleation will become easier and easier as the system is driven far from the equilibrium boundary, whereas growth will be easier at the highest temperature close to the equilibrium boundary. The TTT plots (fig. 8.5) show clearly that there is an incubation time before the transformation becomes noticeable and that it becomes infinite at the Clapeyron (no nucleation) and at too low temperatures (no growth).

For athermal transformations like martensitic transformations (e.g. in steels) the amount of transformed phase does not depend on time but only on the temperature at which the system is cooled (fig. 8.6).

Finally, the case of *non-isothermal transformation kinetics* must be mentioned since it is, in fact, the current case in transformation plasticity: in the general Avrami equation (8.5) the reaction rate is temperature-dependent; now, if the temperature varies with time, so will the reaction rate, and the differential equation (8.4) cannot be integrated as simply.

For a constant heating rate $\dot{T}$, the temperature corresponding to the

Fig. 8.5. Time–Temperature–Transformation curve for a nucleation and growth isothermal transformation. The transformation (for cooling) takes place between the curve $X = \varepsilon$ (vanishingly small quantity of transformed phase) and $X = 1$ (complete transformation).

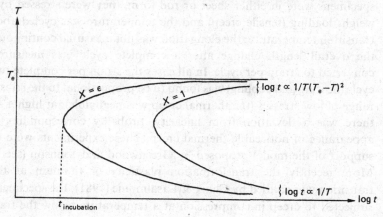

Fig. 8.6. Time–Temperature–Transformation curve for a martensitic transformation. The amount of transformed phase depends only on the final temperature (between M_s and M_f) and does not depend on time.

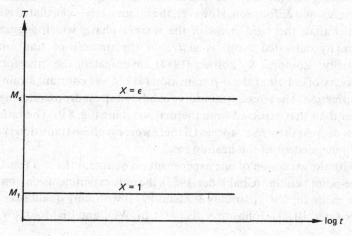

maximum reaction rate can be calculated for a first-order reaction (Meisel & Cote, 1983).

8.3   Experimental evidence for transformation plasticity

After the first results of Sauveur (1924), the transformation plasticity of iron and austenitic steels at the α–γ or martensitic transition was investigated by de Jong & Räthenau (1959, 1961), Oelschlägel & Weiss (1966) and Kot & Weiss (1970) in particular. Greenwood and Johnson (1965) carried out experiments on several metals (Zr, Ti, Co, U, Fe–C); the specimens were in either sheet or rod form, they were stressed by dead-weight loading (tensile creep) and the temperature was cycled about the transition temperature; the elongation was not measured continuously but the overall length change after a complete cycle was measured and converted to strain per cycle. In all cases the strain per complete two-way cycle (heating and cooling) was found to be proportional to the stress in the range of low stresses (the thermal creep was negligible); at higher stresses there was a deviation from linearity, probably corresponding to the appearance of noticeable thermal creep. These experiments were used in support of the model proposed by Greenwood and Johnson (see below). More recently the transformation plasticity of titanium at the α–β transition was studied by Chaix & Lasalmonie (1981); the specimens were subjected to creep in compression at a temperature below the transition point, the temperature was then increased and the strain measured continuously (fig. 8.7). The strain per half-cycle (heating or cooling), connected to the normal thermal creep strain of both phases, was found to be proportional to the applied stress and independent of grain size and of the heating rate (fig. 8.8); the results are in good agreement with the theory of Greenwood & Johnson. However, the internal stress calculated is slightly smaller than the yield stress of the weaker phase, which suggests that thermally activated creep is active in the process of transformation plasticity. Zamora & Poirier (1983) investigated the transformation plasticity of cobalt at the ε–β transition (417 °C) at constant strain-rate in compression. The stress–strain curve exhibits a stress drop (softening) much larger than that expected from thermal softening (fig. 8.9). The ratio of the stress drop to the stress expected if there were no phase transition is $\Delta\sigma/\sigma \approx$ 0.40 independent of the heating rate.

With the exception of one experiment on quartz at the α–β transition in three-point bending (Chaklader, 1963), the only experiments on non-metals were made on CsCl (Sammis & Dein, 1974) with only qualitative results, and oxides: Bi_2O_3 (Johnson *et al.*, 1975), Bi_2WO_6 and Bi_2MoO_6 (Winger *et al.*, 1980) and the eutectoid system Bi_2O_3–Sm_2O_3 (Smyth *et al.*, 1975). The case of Bi_2WO_6 and Bi_2MoO_6 is especially interesting since the specimens (tested in compression creep) exhibited a strain per cycle, linearly dependent on σ, but inversely dependent on grain size (fig. 8.10); heating through the transition at high stresses produced large strains. The strain is therefore produced by a time-dependent, grain-size-sensitive creep mechanism (as in the case of Bi_2O_3, Johnson *et al.*, 1975) which is thought to be grain-boundary sliding. As the strain is continuously measured, the maximum strain-rate is known and the strain-rate sensitivity m can be determined; its high value ($m = 0.85$) confirms that the active mechanism is grain-boundary sliding and that we have here a real case of transformation superplasticity, governed by the internal stresses produced during the transition (see below) and different from structural superplasticity of a fine-grained eutectic produced by the transition; here the superplasticity occurs while the transition is in progress.

8.4 Macroscopic and microscopic models

Greenwood & Johnson (1965) proposed a general macroscopic model for transformation plasticity based on purely mechanical concepts: in a polycrystal undergoing a phase change of the first order, the grains that transform suffer a volume change ΔV, this creates internal stresses between

Fig. 8.7. Enhanced creep during α–β phase transformation in titanium. The creep curve was recorded while temperature was increasing. ε_T is the permanent strain corresponding to transformation plasticity (after Chaix & Lasalmonie, 1981).

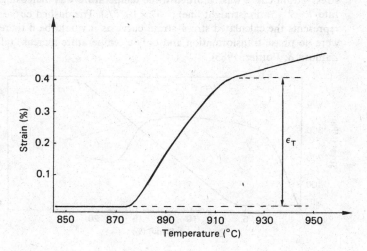

Fig. 8.8. Strain per half-cycle for titanium as a function of stress. Top: $\alpha \rightarrow \beta$ (heating). Bottom: $\beta \rightarrow \alpha$ (cooling) (after Chaix & Lasalmonie, 1981).

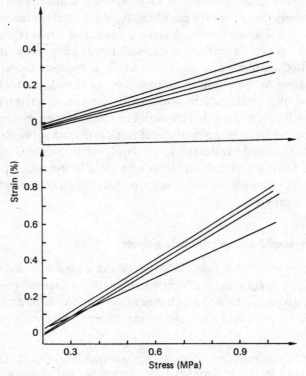

Fig. 8.9. Stress drop during $\varepsilon-\beta$ phase transformation in cobalt. The stress–strain curve was recorded while temperature was increasing at a rate $\dot{T} = 2\,°C/min$ (straight line) ($\dot{\varepsilon} = 2 \times 10^{-5}/s$). The dashed curve represents the calculated stress–strain curve as it would be if there were no phase transformation and only a temperature increase (after Zamora & Poirier, 1983).

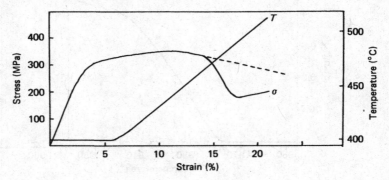

Fig. 8.10. Transformation plasticity in Bi_2WO_6. (*a*) Strain per cycle as a function of stress for two grain sizes. (*b*) Strain-rate sensitivity *m* for 3 μm grain size (after Winger *et al.*, 1980).

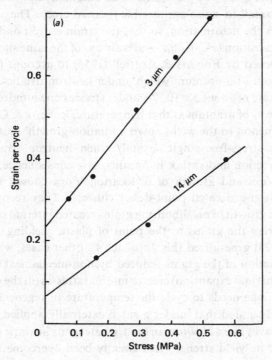

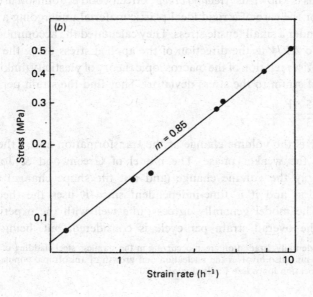

grains, when they reach the value of the yield stress (elastic limit) of the weaker phase the polycrystal behaves as though its elastic limit were vanishingly small and gives way plastically to an externally applied stress that would be too small to cause appreciable thermal creep. The applied stress merely guides the deformation, so that the strain corresponding to both senses of the transition ($\alpha \rightarrow \beta$ or $\beta \rightarrow \alpha$) is always of the same sign. This idea was first proposed by Roberts & Cottrell (1956) to account for the remarkable behaviour of α-uranium which, under neutron irradiation, at 100 °C, creeps at a rate of about 3×10^{-4}/s under stresses one hundredth of the normal yield stress of uranium at that temperature. Roberts & Cottrell related this phenomenon to the well-known 'radiation growth' that takes place in uranium: stress-free single crystals when neutron irradiated lengthen in one direction and shrink in another as a consequence of the anisotropic nucleation and growth of dislocation loops caused by the collapse of irradiation-generated point-defect clusters.[†] They recognized that the anisotropic growth of neighbouring grains created internal stresses large enough to bring the grains to the point of plastic yielding. Later, Cottrell (1964, p. 338) generalized this approach to other cases, when an anisotropic deformation of the grains induced by non-mechanical causes (like, for instance, thermal expansion) creates internal stresses (in the case of thermal expansion, one needs to cycle the temperature to regenerate the internal stresses). He stated that 'under a small externally applied stress, such a material behaves like a Newtonian fluid, whatever its intrinsic plastic properties, because its yield strength has already been overcome by the stresses'. This is known as '*yielding creep*'. Greenwood & Johnson used this approach for the case of a rigid ideal plastic material undergoing a phase transition under a small tensile stress. They calculated the accommodation strain due to $\Delta V/V$ in the direction of the applied stress using the linear Levy–von Mises relation of the macroscopic theory of plasticity (linking the increment of strain to the stress deviator). They find the strain per cycle:

$$\varepsilon = \frac{5}{3} \frac{\Delta V}{V} \frac{\sigma}{Y} \tag{8.15}$$

where $\Delta V/V$ is the volume change at the transformation and Y the yield strength of the weaker phase. The model of Greenwood & Johnson considers only the volume change (and not the shape change) at the transformation and it is time-independent since it uses the theory of plasticity. The model generally agrees quite well with the experiments whenever the overall strain per cycle is considered, but, being time-

[†] Incidentally, irradiation creep occurring in the stainless steel cladding of nuclear fuel pins also involves the nucleation and growth of anisotropic populations of dislocation loops (see Caillard *et al.*, 1980).

independent, it obviously is inadequate to account for the creep-rate enhancement during the phase transition. Kot & Weiss (1970) proposed a microscopic model in which they consider that the internal stress σ_T due to the transformation is added to the applied stress and drives the dislocations faster; assuming a dependence of the dislocation velocity on stress of the form:

$$v = v_s \exp\left(-\frac{B}{\sigma}\right)$$

the resulting creep-rate is

$$\dot{\varepsilon} = \rho b v_s \exp\left(-\frac{B}{\sigma + \sigma_T}\right) \tag{8.16}$$

If $\sigma \ll \sigma_T$ the expression can be linearized in σ. Kot & Weiss finally find an expression for the strain per cycle similar to that of Greenwood & Johnson.

In opposition to these models in which the internal stress is unrelaxed and does not relax during creep, Poirier (1982) proposed a microscopic model where the total transformation strain (shape and volume change) is relaxed and accommodated by an extra density of dislocations which can then move under the applied stress and create an extra creep-rate. The model need not assume that the thermal creep is small, and it takes into account the kinetics of the phase transition. For the simple case considered by Greenwood & Johnson (zero thermal creep and transformation strain purely dilatational) the transformation creep rate reduces to:

$$\dot{\varepsilon} = C \frac{\Delta V}{V} \sigma X (1 - X) \tag{8.17}$$

where $X(t)$ is the volume fraction of transformed phase. For a first-order kinetics, the strain per half-cycle can be calculated by integration:

$$\varepsilon = \frac{C}{2K} \frac{\Delta V}{V} \sigma \tag{8.18}$$

The expression (8.18) is linear in σ and $\Delta V/V$ like the one found by Greenwood & Johnson, but it also depends on the reaction rate $K(T)$.

Paterson (1983) improved this model by taking recovery into account, thus introducing a lifetime for the extra dislocations created to accommodate the internal stresses. The resulting creep-rate thus depends on the kinetics of phase transformation (creation of dislocations), the kinetics of deformation (motion of dislocations) and the kinetics of recovery (annihilation of dislocations). Finally, extra dislocations may also be produced as a consequence of the frequently observed drop in shear modulus just before a phase transition. As the critical length for the activation of a Frank–Read source varies as μ/σ, a drop in modulus may cause shorter dislocation segments to act as sources under the same stress, thus increasing the dislocation density and causing an extra creep-rate (Poirier, 1982).

Readings

A. Putnis & J. D. C. McConnell, *Principles of mineral behaviour*, Elsevier, New York (1980), 257 pp. A very good review of the thermodynamics and kinetics of phase transformations and application to minerals.
C. N. R. Rao & K. J. Rao, *Phase transitions in solids*, McGraw-Hill (1978), 330 pp. A clear and complete textbook on all phase transitions in solids.

9

Scaling and classification

After examining the various high-temperature deformation mechanisms in the preceding chapters, it is proper to step back and take a look at all the possible deformation mechanisms for a given material according to mechanical criteria. This is a considerable job, but it has recently been done for us in a book by Frost & Ashby (1982). Our last chapter will therefore be nothing more than a sketchy summary of the subject matter of their book.

9.1 Scaling

Prior to any attempt to compare and classify objects or behaviours, one must make sure to compare only comparable quantities. We have already seen that a temperature of 100 °C has not the same meaning for indium, which melts at 157 °C, and magnesium oxide, which melts at 2852 °C. It follows that *the comparison of the deformation behaviour of materials can only be made on the basis of properly scaled, dimensionless quantities* (hydrodynamicists proceed in the same way).

The obvious and widely used normalizing parameters for temperature and stress are the absolute melting temperature and some elastic modulus (preferably μ), so that the dimensionless parameters are T/T_m and σ/μ. However, this choice is not the only one and it may not even be the best (Frost & Ashby, 1982): energies per molar volume have the dimension of stresses, and energies over the Boltzmann constant k have the dimension of temperature. The criterion for a good scaling parameter is the possibility of shrinking the cloud of data points onto a single curve with a minimum scatter. Then the best scaling parameter for temperature is $\mu\Omega/k$ and the best one for stresses is the cohesive energy per molar volume $\Delta H_c/\Omega$. However, for practical purposes T_m and μ are more convenient. Strain rates can be normalized using $D_{T_m}/\Omega^{2/3}$, where D_{T_m} is the diffusion coefficient at the melting temperature. It is interesting to attempt a classification of various materials according to their mechanical properties. Frost & Ashby define *isomechanical groups*: 'The members of a group are mechanically

Fig. 9.1. Isomechanical groups. (*a*) Materials classed according to the value of dimensionless parameter Q_v/RT_m (Q_v = activation energy for volume diffusion, T_m = melting point in degrees K). (*b*) Materials classed according to the value of dimensionless parameter $\mu_0\Omega/kT_m$. μ_0 shear modulus at RT. Ω atomic or molar volume. T_m melting point in degrees K (Frost & Ashby, 1982).

(*a*)

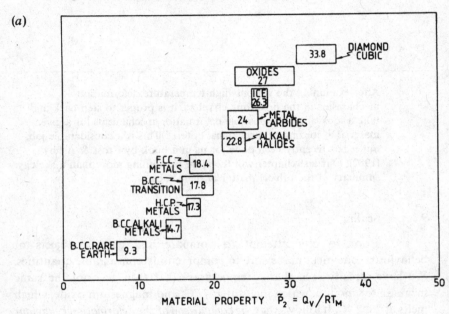

(*b*)

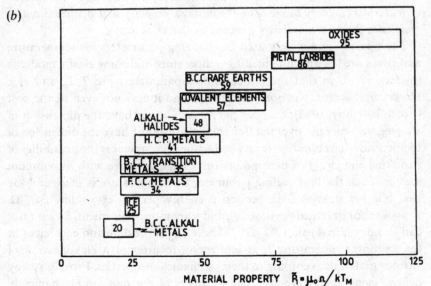

similar and by proper scaling, their mechanical properties can be brought more or less into coincidence. The same scaling separates and distinguishes this group from other groups'.

Good dimensionless material properties used to identify the isomechanical groups are $\mu\Omega/kT_m$ (fig. 9.1) and Q_v/RT_m where Q_v is the activation energy for volume diffusion.

In this way the division into groups takes into account not only the structure but also the bonding: crystals with the rock-salt structure, for instance, can be divided into four groups: alkali halides (NaCl, LiF, AgCl, ...), simple oxides (MgO, FeO, ...), lead sulphide (PbS, PbTe, ...), metal carbides (TiC, VC, ...). The notion of isomechanical group is of the highest relevance for geophysics: since most of the mantle materials (except olivine) are high-pressure materials synthesized in very small quantities, the only

Fig. 9.2. Deformation-mechanism maps for various materials – grain size 100 μm. (*a*) Pure iron. (*b*) Pure silicon. (*c*) MgO. (*d*) Ice I$_h$ (Frost & Ashby, 1982).

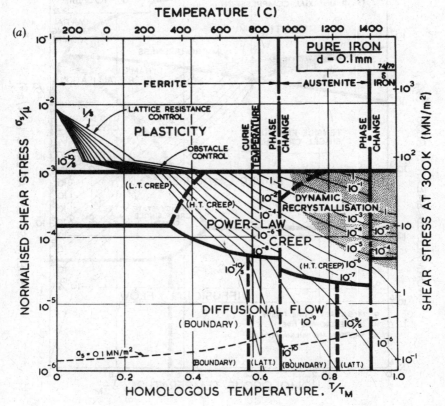

way, so far, to obtain some information on their viscosity is to investigate *analogue* materials, chosen on the basis of a common structure (which ensures that dislocations can be described in the same way) and, if possible, with the same chemical bonding. The information one can obtain in this manner can obviously be extrapolated to the actual material with a higher degree of confidence if the relations between isomechanical groups are better known. For instance, the high-temperature behaviour of the crystals with perovskite structure is still unknown except for preliminary experiments on $KZnF_3$ (Poirier *et al.*, 1983); fluoride perovskites presumably form an isomechanical group, and so do, presumably, the oxide perovskites ($BaTiO_3$, $MgSiO_3$, ...) but are these groups related in the same way as the groups for crystals with fluorite structure (CaF_2, BaF_2, ThO_2, CaO_2)?

Fig. 9.2(*b*)

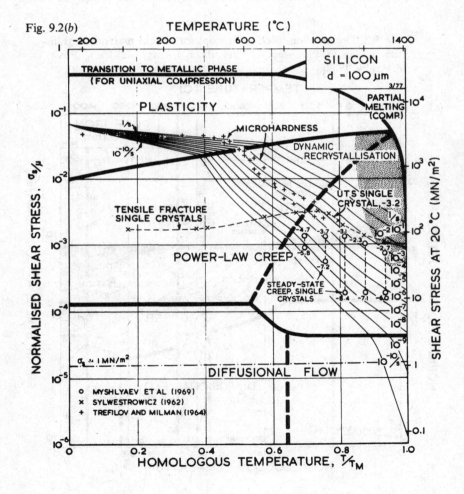

9.2 Deformation maps

The mechanical behaviour of each material of a group must then be characterized for all the possible physical conditions; the information can be condensed in *deformation-mechanism maps* first proposed by Ashby (1972). The idea consists in describing the mechanical behaviour of the material in terms of constitutive equations fitted to experimental data and representing the equations by a surface in a multidimensional-variable space:

$$f(\dot{\varepsilon}, \sigma, T, P, p_{O_2}, \ldots) = 0 \qquad (9.1)$$

Any variable can be singled out to be expressed as a function of the others.

The surface is constituted of several portions, each corresponding to a dominant mechanism (e.g. giving the highest $\dot{\varepsilon}$) and separated by 'curves' determined by equating two constitutive laws. The surfaces can be conveniently projected on any one of the coordinate planes, and contoured

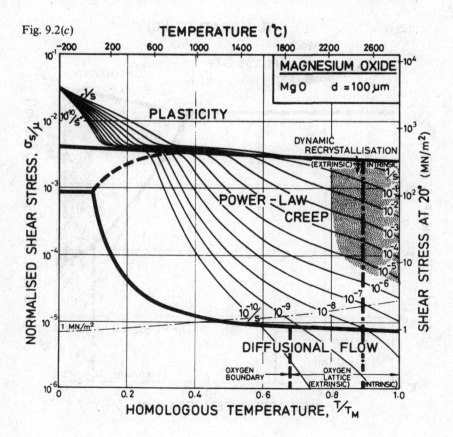

Fig. 9.2(c)

for definite values of another variable, to give *deformation maps*. The most common choice is to use projections on the plane of normalized temperature and stress (T/T_m, σ/μ) with strain-rate contours. All the other parameters have to be kept constant (grain size for instance). Grain size d can be used in deformation maps, but only if it is an independent variable. Examples of deformation maps for materials of various isomechanical groups are given in fig. 9.2 (Frost & Ashby, 1982).

Deformation maps are extremely useful as a means to obtain synthetic information on the overall mechanical behaviour of a material (including low-temperature deformation or fracture), as a guide for extrapolation, for instance to geophysical conditions. However, as pointed out by Frost & Ashby, it must be recognized that 'the maps are no better (and no worse) than the equations and data used to construct them'. Caution must therefore be exerted in extrapolating towards domains where absolutely no experimental data exist.

Fig. 9.2(*d*)

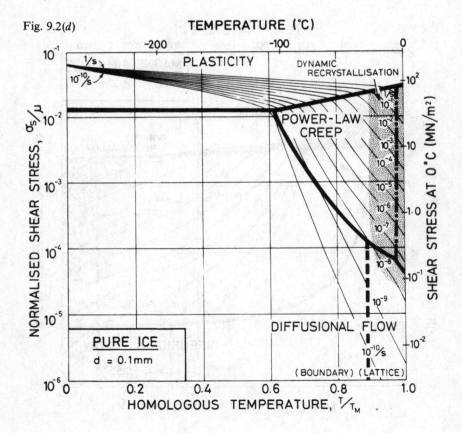

The present book could therefore end with a plea to experimentalists in materials science to publish their data in an untreated form, so that more reliable deformation maps for more materials can be constructed, as physical models and constitutive laws are perfected.

Readings

H. J. Frost & M. F. Ashby, *Deformation-mechanism maps*, Pergamon Press (1982), 166 pp. Obviously!

References

Ahlquist, C. N., & Nix, W. D. (1969). A technique for measuring mean internal stress during high temperature creep. *Scripta Metall.*, **3**, 679–82.

Aladag, E., Davis, L. A., & Gordon, R. B. (1970). Cross-slip and plastic deformation of NaCl single and polycrystals at high pressure. *Philos. Mag.*, **21**, 469–78.

Alexander, H., & Haasen, P. (1968). Dislocation and plastic flow in the diamond structure. *Solid State Phys.*, **22**, 27–158.

Anderson, O. L. (1982). The Earth's core and the phase diagram of iron. *Phil. Trans. R. Soc. London*, A.**306**, 21–35.

Andrade, E. N. da C. (1910). On the viscous flow in metals and allied phenomena. *Proc. Roy. Soc. London*, A.**84**, 1–12.

Andrade, E. N. da C. (1914). The flow of metals under large constant stresses. *Proc. Roy. Soc. London*, A.**90**, 329–42.

Argon, A. S., & Moffatt, W. C. (1981). Climb of extended edge dislocations. *Acta Metall.*, **29**, 293–99.

Ashby, M. F. (1969). On interface-reaction control of Nabarro–Herring creep and sintering. *Scripta Metall.*, **3**, 837–42.

Ashby, M. F. (1970). The deformation of plastically non-homogeneous crystals. *Philos. Mag.*, **21**, 399–424.

Ashby, M. F. (1972). A first report on deformation-mechanism maps. *Acta Metall.*, **20**, 887–97.

Ashby, M. F., Spaepen, F., & Williams, S. (1978). The structure of grain boundaries described as a packing of polyhedra. *Acta Metall.*, **26**, 1647–63.

Ashby, M. F., & Verrall, R. A. (1973). Diffusion accommodated flow and superplasticity. *Acta Metall.*, **21**, 149–63.

Ashby, M. F., & Verrall, R. A. (1978). Micromechanisms of flow and fracture and their relevance to the rheology of the upper mantle. *Phil. Trans. Roy. Soc. London*, **288A**, 59–95.

Auten, T. A., Davis, L. A., & Gordon, R. B. (1973). Hydrostatic pressure and the mechanical properties of NaCl polycrystals. *Philos. Mag.*, **28**, 335–41.

Auten, T. A., Radcliffe, S. V., & Gordon, R. B. (1976). Flow stress of MgO simple crystals compressed along [100] at high hydrostatic pressure. *J. Amer. Ceram. Soc.*, **59**, 40–2.

Avery, D. H., & Backofen, W. A. (1965). A structural basis for superplasticity. *Trans. ASM*, **58**, 551–62.

Ayensu, A., & Ashbee, K. H. G. (1977). The creep of quartz simple crystals, with special reference to the mechanism by which water accommodates dislocation glide. *Philos. Mag.*, **36**, 713–23.

Backofen, W. A., Turner, I. R., & Avery, D. H. (1964). Superplasticity in an Al–Zn alloy. *Trans. ASM*, **57**, 980–90.

Bailey, J. E., & Hirsch, P. B. (1962). The recrystallization process in some polycrystalline metals. *Proc. Roy. Soc. London*, **A267**, 11–30.

Balasubramanian, N., & Li, J. C. M. (1970). The activation area for creep deformation. *J. Mater. Sci.*, **5**, 434–44.

Balluffi, R. W. (1970). On measurements of self-diffusion rates along dislocations in FCC metals. *Phys. Stat. Solid.*, **42**, 11–34.

Balluffi, R. W., Bristowe, P. D., & Sun, C. P. (1981). Structure of high-angle grain boundaries in metals and ceramic oxides. *J. Amer. Ceram. Soc.*, **64**, 23–30.

Balluffi, R. W., Brokman, A., & King, A. H. (1982). CSL/DSC lattice model for general crystal–crystal boundaries and their line defects. *Acta Metall.*, **30**, 1453–70.

Balluffi, R. W., & Cahn, J. W. (1981). Mechanism for diffusion-induced grain boundary migration. *Acta Metall.*, **29**, 493–500.

Barrett, C. R., Ardell, A. J., & Sherby, O. D. (1964). Influence of modulus on the temperature dependence of the activation energy for creep at high temperature. *Trans. AIME*, **230**, 200–4.

Barrett, C. R., & Sherby, O. D. (1965). Influence of stacking fault energy on high temperature creep of pure metals. *Trans. AIME*, **233**, 1116–19.

Bassett, W. A. (1979). The diamond cell and the nature of the Earth's mantle. *Ann. Rev. Earth Planet. Sci.*, **7**, 357–84.

Bedère, D., Jamard, C., Jarlaud, A., & François, D. (1970). Deformation of beryllium single crystals under hydrostatic pressure. *Phys. Stat. Sol. (a)*, **1**, 135–42.

Bell, P. M. (1979). Ultra high pressure experimental mantle mineralogy. *Rev. Geophys. Space Phys.*, **17**, 788–91.

Bellier, S. P., & Doherty, R. D. (1977). The structure of deformed aluminium and its recrystallization. Investigations with transmission Kossel diffraction. *Acta Metall.*, **25**, 521–38.

Belzner, V., & Granzer, F. (1977). Atomistic calculations on the dissociation of an edge dislocation in NaCl. *Phys. Stat. Solid. (a)*, **39**, 183–87.

Bergman, B. (1975). A note on 'the strain transient dip test'. *Scand. J. Metallurgy*, **4**, 109–12.

Beyeler, M., & Adda, Y. (1968). Détermination des volumes d'activation pour la diffusion des atomes dans l'or, le cuivre et l'aluminium. *J. Physique*, **29**, 345–52.

Birch, J. M., & Wilshire, B. (1974). The effect of stress changes during creep of single and polycrystalline MgO. *Philos. Mag.*, **30**, 1023–31.

Blacic, J. D. (1975). Plastic deformation mechanisms in quartz: the effect of water. *Tectonophysics*, **27**, 271–94.

Blacic, J. D. (1981). Water diffusion in quartz at high pressure: tectonic implications. *Geophys. Res. Lett.*, **8**, 721–3.

Bollmann, W. (1970). *Crystal defects and crystalline interfaces.* Springer-Verlag, New York, 254 pp.

Boullier, A. M., & Gueguen, Y. (1975). Origin of some mylonites by superplastic flow. *Contrib. Mineral. Petrol.*, **50**, 93–104.

Brebec, G., & Poirier, J. P. (1975). Fluage des solutions solides concentrées binaires contrôlé par la diffusion: choix d'un coefficient de diffusion. *Acta Metall.*, **23**, 125–30.

Bretheau, T., Castaing, J., Rabier, J., & Veyssière, P. (1979). Mouvement des dislocations et plasticité à haute température des oxydes binaires et ternaires. *Adv. in Phys.*, **28**, 835–1014.

Bretheau, T., & Dolin, C. (1978). Heterogeneous deformation of Cu_2O single crystals during high temperature compression creep. *J. Mater. Sci.*, **13**, 587–93.

Bretheau, T., Pelissier, B., & Sieber, B. (1981). Plastic properties of Cu_2O, mechanical tests and transmission electron microscopy – II. High temperature. *Acta Metall.*, **29**, 1617–29.

Bristowe, P. D., & Sass, S. L. (1980). The atomic structure of a large angle [001] twist boundary in gold determined by a joint computer modelling and X-Ray diffraction study. *Acta Metall.*, **28**, 575–88.

Brown, A. M., & Ashby, M. F. (1980). On the power-law creep equation. *Scripta Metall.*, **14**, 1297–1302.

Bullen, F. P., Henderson, F., Wain, H. L., & Paterson, M. S. (1964). The effect of hydrostatic pressure on brittleness in chromium. *Philos. Mag.*, **9**, 803–15.

Burton, B. (1971). A 'creep yield stress' for superplastic deformation. *Scripta Metall.*, **5**, 669–72.

Burton, B. (1977). *Diffusional creep of polycrystalline materials*. Diffusion & defect monograph series. Trans. Tech. Pub., Aedermansdorff, 119 pp.

Burton, B. (1982a). The dislocation network theory of creep. *Philos. Mag.*, A.**45**, 657–75.

Burton, B. (1982b). The influence of solute drag on dislocation creep. *Philos. Mag.*, A.**46**, 607–16.

Burton, B. (1982c). The influence of stacking fault energy on creep. *Acta Metall.*, **30**, 905–10.

Butcher, B. M., & Ruoff, A. L. (1961). Effect of hydrostatic pressure on the high temperature steady-state creep of lead. *J. Appl. Phys.*, **32**, 2036–38.

Cahn, J. W. (1956). The kinetics of grain boundary nucleated reactions. *Acta Metall.*, **4**, 449–59.

Caillard, D., & Martin, J. L. (1982, 1983). Microstructure of aluminium during creep at intermediate temperature. I: Dislocation between networks after creep. *Acta Metall.*, **30**, 437–45. II: *In situ* studies of sub-boundary properties. *Acta Metall.*, **30**, 791–8. III: The rate controlling process. *Acta Metall.*, **31**, 813–25.

Caillard, D., Martin, J. L., & Jouffrey, B. (1980). Creep under irradiation of 316 steel in the high voltage electron microscope. *Acta Metall.*, **28**, 1059–67.

Cannon, W. R., & Sherby, O. D. (1970a). High temperature creep behaviour of class I and class II solid-solution alloys. *Metall. Trans.*, **1**, 1030–2.

Cannon, W. R., & Sherby, O. D. (1970b). High temperature creep of NaCl–KCl solid solution alloys. *J. Amer. Ceram. Soc.*, **53**, 346–9.

Cannon, W. R., & Sherby, O. D. (1973). Third-power stress dependence in creep of polycrystalline nonmetals. *J. Amer. Ceram. Soc.*, **56**, 157–60.

Chaix, C., & Lasalmonie, A. (1981). Transformation induced plasticity in titanium. *Res. Mechanica*, **2**, 241–9.

Chaklader, A. C. D. (1963). Deformation of quartz crystals at the transformation temperature. *Nature*, **197**, 791–2.

Chen, W. K., & Peterson, N. L. (1980). Iron diffusion and electrical conductivity in magnesio-wüstite solid solutions (Mg, Fe)O. *J. Phys. Chem. Solids*, **41**, 335–9.

Chevalier, G. T., McCormick, P., & Ruoff, A. L. (1967). Pressure dependence of high-temperature creep in single crystals of indium. *J. Appl. Phys.*, **38**, 3697–700.

Chevenard, P. (1922). Alliages de Nickel conservant leur rigidité dans un domaine étendu de température. *C.R. Acad. Sci. Paris*, **175**, 486–9.

Chongmo, L., & Hillert, M. (1982). Diffusion induced grain boundary migration in Cu–Zn. *Acta Metall.*, **30**, 1133–45.

Christie, J., & Ord, A. (1980). Flow stress from microstructure of mylonites: example and current assessment. *J. Geophys. Res.*, **85**, 6253–62.

Christie, J. M., Ord, A., & Koch, P. S. (1980). Relationship between recrystallized grain size and flow stress in experimentally deformed quartzite. *EOS*, **61**, 377.

Chu, S. N. G., & Li, J. C. M. (1977). Impression creep: a new creep test. *J. Mater. Sci.*, **12**, 2200–8.

Chua, J. O., & Ruoff, A. L. (1975). Pressure dependence of the yield stress of potassium at low homologous temperatures. *J. Appl. Phys.*, **46**, 4659–63.

Chung, T. E., & Davies, T. J. (1979). The low stress creep of fine-grain uranium dioxide. *Acta Metall.*, **27**, 627–35.

Cicotti, G., Guillopé, M., & Pontikis, V. (1983). High angle grain boundary premelting transition: a molecular dynamics study. *Phys. Rev.*, B.27, 5576–85.

Clauer, A. H., Wilcox, B. A., & Hirth, J. P. (1970). Dislocation substructure induced by creep in molybdenum single crystals. *Acta Metall.*, **18**, 381–97.

Coath, J. A., & Wilshire, B. (1977). Deformation processes during high temperature creep of Lime, Magnesia and Doloma. *Ceramurgia International*, **3**, 103–8.

Coble, R. L. (1963). A model for boundary-diffusion controlled creep in polycrystalline materials. *J. Appl. Phys.*, **34**, 1679–82.

Cottrell, A. H. (1964). *The mechanical properties of matter*. Wiley, New York, 430 pp.

Cox, S. F., Etheridge, M. A., & Hobbs, B. E. (1981). The experimental ductile deformation of polycrystalline and single crystal pyrite. *Econ. Geol.*, **76**, 2105.

Crossland, I. G. (1974). Low stress creep of cadmium. *Phys. Stat. Sol. (a)*, **23**, 231–5.

Crossland, I. G., Jones, R. B., & Lewthwaite, G. W. (1973). The use of helically coiled springs in creep experiments with special reference to the case of Bingham flow. *J. Phys. D.*, **6**, 1040–6.

Darot, M., & Gueguen, Y. (1981). High-temperature creep of forsterite single-crystals. *J. Geophys. Res.*, **86**, 6219–34.

Davidge, R. W., & Pratt, P. L. (1964). Plastic deformation and work-hardening in NaCl. *Phys. Stat. Sol.*, **6**, 759–76.

Davies, G. F. (1977). Whole mantle convection and plate tectonics. *Geophys. J. R. Astr. Soc.*, **49**, 459–86.

Davis, L. A., & Gordon, R. B. (1968). Pressure dependence of the plastic flow stress of alkali halide single crystals. *J. Appl. Phys.*, **39**, 3885–97.

Deeley, R. M. (1895). Glacier grains. *Philos. Mag.*, **39**, 453–5.

de Jong, M., & Räthenau, G. W. (1959). Mechanical properties of iron and some iron alloys while undergoing allotropic transformation. *Acta Metall.*, **7**, 246–53.

de Jong, M., & Räthenau, G. W. (1961). Mechanical properties of an iron carbon alloy during allotropic transformation. *Acta Metall.*, **9**, 714–20.

De Vries, K. L., & Gibbs, P. (1963). Pressure dependence of creep in Zn and Cd *J. Appl. Phys.*, **34**, 3119–20.

De Vries, K. L., Baker, G. S., & Gibbs, P. (1963a). Pressure dependence of the creep of lead. *J. Appl. Phys.*, **34**, 2254–7.

De Vries, K. L., Baker, G. S., & Gibbs, P. (1963b). Effect of pressure on creep in tin. *J. Appl. Phys.*, **34**, 2258–9.

Dhalenne, G., Dechamps, M., & Revcolevschi, A. (1982). Relative energies of $\langle 011 \rangle$ tilt boundaries in NiO. *J. Amer. Ceram. Soc.*, **65**, 611–12.

Dixon-Stubbs, P. J., & Wilshire, B. (1982). Deformation processes during creep of single and polycrystalline CaO. *Philos. Mag.*, A.**45**, 519–29.

Donlon, W. T., Mitchell, T. E., & Heuer, A. H. (1979). Climb dissociation of network dislocations in non-stoichiometric Mg–Al spinel. *Philos. Mag.*, A.**40**, 351–66.

Dorn, J. E. (1956). Some fundamental experiments on high-temperature creep. Creep and fracture of metals at high temperatures. In *NPL symposium, 1954*, HMSO, London, 89–138.

Dorn, J. E., & Jaffe, N. (1961). Effect of temperature on the creep of polycrystalline aluminium by the cross-slip mechanism. *Trans. AIME*, **221**, 229–33.

Doukhan, J. C., & Paquet, J. (1982). Plastic deformation of andalusite single crystal Al_2SiO_5. *Bull. Minéral.*, **105**, 170–5.

Doukhan, N., Duclos, R., & Escaig, B. (1973). Structural and mechanical study of creep in Al_2O_4Mg single crystal. *J. Physique*, **34** (C.9), 379–87.

Doukhan, N., Duclos, R., & Escaig, B. (1979). Sessile dissociation in the stoichiometric spinel $MgAl_2O_4$. *J. Physique Paris*, **40**, 381–7.

Doukhan, N., Duclos, R., & Escaig, B. (1982). Climb dissociation in {113} planes in Al–Mg spinels. *J. Physique*, **43**, 1149–57.

Duclos, R. (1981). High-temperature deformation of $MgO.1.8\ Al_2O_3$ spinel single crystals of $\langle 111 \rangle$ axial orientation. *J. Phys. Paris*, **42**, 49–57.

Duclos, R., Doukhan, N., & Escaig, B. (1978). High-temperature creep behaviour of nearly stoichiometric alumina spinel. *J. Mater. Sci.*, **13**, 1740–8.

Duclos, R., Doukhan, N., & Escaig, B. (1982). Study of the origin of the composition influence on the mechanical properties of MgO, nAl_2O_3 spinels. *Acta Metall.*, **30**, 1381–8.

Durham, W. B., Froidevaux, C., & Jaoul, O. (1979). Transient and steady-state creep of pure forsterite at low stress. *Phys. Earth Planet. Interiors*, **19**, 263–74.

Durham, W. B., & Goetze, C. (1977a). Plastic flow of oriented single crystals of olivine. 1. Mechanical data. *J. Geophys. Res.*, **82**, 5737–53.

Durham, W. B., & Goetze, C. (1977b). A comparison of the creep properties of pure forsterite and iron bearing olivine. *Tectonophysics*, **40**, T15–T18.

Durham, W. B., Goetze, C., & Blake, B. (1977). Plastic flow of oriented single crystals of olivine. 2. Observations and interpretations of the dislocation structure. *J. Geophys. Res.*, **82**, 5755–70.

Durham, W. B., Heard, H. C., & Kirby, S. H. (1984). Experimental deformation of polycrystalline H_2O ice at high pressure and low temperature. Preliminary results. *J. Geophys. Res.* (in press).

Duval, P., Ashby, M. F., & Anderman, I. (1983). Rate controlling processes in the creep of polycrystalline ice. *J. Phys. Chem.*, **87**, 4066–74.

Dziewonski, A. M., Hales, A. L., & Lapwood, E. R. (1975). Parametrically single earth models consistent with geophysical data. *Phys. Earth Planet. Int.*, **10**, 12–48.

Edelin, G., & Poirier, J. P. (1973). Etude de la montée des dislocations au moyen d'expériences de fluage par diffusion dans le magnésium. I. Mécanisme de déformation. *Philos. Mag.*, **28**, 1203–10.

Edington, J. W., Melton, K. N., & Cutler, C. P. (1976). Superplasticity. *Progr. Mater. Sci.*, **21**, 61–70.

Elsasser, W. M., Olson, P., & Marsh, B. D. (1979). The depth of mantle convection. *J. Geophys. Res.*, **84**, 147–55.

Escaig, B. (1967). Sur l'origine de la limite élastique des métaux cubiques centrés à basse température. *J. Physique*, **28**, 171–86.

Escaig, B. (1968a). Sur le glissement dévié des dislocations dans la structure cubique à faces centrées. *J. Physique*, **29**, 225–39.

Escaig, B. (1968b). L'activation thermique des déviations sous faibles contraintes dans les structures hc et cc. *Phys. Stat. Sol.*, **28**, 463–74.

Etheridge, M. A. (1975). Deformation and recrystallization of orthopyroxene from the Giles complex, Central Australia. *Tectonophysics*, **25**, 87–114.

Etheridge, M. A., & Hobbs, B. E. (1974). Chemical and deformational controls on recrystallization of mica. *Contrib. Mineral. Petrol.*, **43**, 111–24.

Etheridge, M. A., & Wilkie, J. C. (1979). Grain size reduction, grain boundary sliding and the flow strength of mylonites. *Tectonophysics*, **58**, 159–78.

Evans, A. G., & Rawlings, R. D. (1969). The thermally activated deformation of crystalline materials. *Phys. Stat. Sol.*, **34**, 9–31.

Exell, S. F., & Warrington, D. H. (1972). Subgrain boundary migration in aluminium. *Philos. Mag.*, **26**, 1121–36.

Eyring, H. (1936). Viscosity, plasticity, and diffusion as examples of absolute reaction rates. *J. Chem. Phys.*, **4**, 283–91.

Fletcher, F. B., Cohen, M., & Hirth, J. P. (1974). Temperature dependence of the strength-differential effect in hardened steels. *Met. Trans. Sci.*, **21**, 61–170.

Fontaine, G. (1968). Dissociation des dislocations sur les plans (110) dans les cristaux ioniques du type NaCl. *J. Phys. Chem. Sol.*, **29**, 209–14.

Fontaine, G., & Haasen, P. (1969). Hydrostatic pressure and plastic deformation of alkali halides. *Phys. Stat. Sol.*, **31**, K67–K70.

Freer, R. (1981). Diffusion in silicate minerals and glasses: A data digest and guide to the literature. *Contrib. Mineral. Petrol.*, **76**, 440–54.

Friedel, J. (1959). Dislocation interactions and internal strains. In *Internal stresses and fatigue in metals*, G. M. Rassweiler & W. L. Grube Eds. Elsevier, Amsterdam, 220–62.

Friedel, J. (1964). *Dislocations*. Pergamon Press, Oxford, 491 pp.

Friedel, J. (1977). Sur le fluage par déviation. *Rev. Physique Appl.*, **12**, 1649–54.

Frost, H. J., & Ashby, M. F. (1982). *Deformation-mechanism maps*. Pergamon Press, Oxford, 166 pp.

Gaboriaud, R. J. (1981). Fluage haute température du sesquioxyde d'Yttrium: Y_2O_3. *Philos. Mag.*, A.**44**, 561–87.

Garofalo, F. (1963). An empirical relation defining the stress dependence of minimum creep rate in metals. *Trans. AIME*, **227**, 351.

Gelles, S. H. (1966). Hydrostatic pressure induced deformation of polycrystalline zinc. *Trans. AIME*, **236**, 981–7.

Ghosh, A. K., & Raj, R. (1981). Grain size distribution effects in superplasticity. *Acta Metall.*, **29**, 607–16.

Gifkins, R. C. (1970). Transitions in creep behaviour. *J. Mater. Sci.*, **5**, 156–65.

Gifkins, R. C. (1976). Grain boundary sliding and its accommodation during creep and superplasticity. *Metall. Trans.*, **7A**, 1225–32.

Gifkins, R. C. (1978). Grain rearrangements during superplastic deformation. *J. Mater. Sci.*, **13**, 1926–36.

Gifkins, R. C., & Langdon, T. G. (1978). Comments on theories of superplasticity. *Mater. Sci. & Eng.*, **36**, 27–33.

Glen, J. W. (1968). The effect of hydrogen disorder on dislocation movement and plastic deformation of ice. *Phys. Kondens. Mater.*, **7**, 43–51.

Glover, G., & Sellars, C. M. (1973). Recovery and recrystallization during high temperature deformation of α-iron. *Metall. Trans.*, **4**, 765–75.

Goodman, D. J., Frost, H. J., & Ashby, M. F. (1981). The plasticity of polycrystalline ice. *Philos. Mag.*, A.**43**, 655–95.

Gordon, R. B. (1965). Diffusion creep in the Earth's mantle. *J. Geophys. Res.*, **70**, 2413–18.

Gordon, R. B. (1971). Observation of crystal plasticity under high pressure with application to the Earth's mantle. *J. Geophys. Res.*, **76**, 1248–54.

Gottstein, G., & Kocks, U. F. (1983). Dynamic recrystallization and dynamic recovery in ⟨111⟩ single crystals of nickel and copper. *Acta Metall.*, **31**, 175–88.

Greenwood, G. W., & Johnson, R. H. (1965). The deformation of metals under small stresses during phase transformations. *Proc. R. Soc. Lond.*, A.**283**, 403–22.

Griggs, D. T. (1967). Hydrolytic weakening of quartz and other silicates. *Geophys. J. R. Astr. Soc.*, **14**, 19–31.

Griggs, D. (1974). A model of hydrolytic weakening in quartz. *J. Geophys. Res.*, **79**, 1653–61.

Griggs, D. T., & Blacic, J. D. (1965). Quartz: anomalous weakness of synthetic crystals. *Science*, **147**, 292–5.

Gronsky, R. (1979). Direct imaging of grain boundaries. In *Grain boundary structure and kinetics*. ASM, Metals Park, Ohio, 470 pp.

Gueguen, Y., & Darot, M. (1982). Les dislocations dans la forstérite déformée à haute température. *Philos. Mag.*, A.**45**, 419–42.

Gueguen, Y., & Nicolas, A. (1980). Mantle rocks deformation. *Ann. Rev. Earth Planet. Sci.*, **8**, 119–44.

Guillopé, M. (1981). Contribution à l'étude de la recristallisation dynamique. Sc.D. Thesis, Univ. Paris 6.

Guillopé, M., & Poirier, J. P. (1979). Dynamic recrystallization during creep of single-crystalline halite: an experimental study. *J. Geophys. Res.*, **84**, 5557–67.

Guillopé, M., & Poirier, J. P. (1980). A model for stress-induced migration of tilt grain boundaries in crystals of NaCl structure. *Acta Metall.*, **28**, 163–7.

Guiu, F., & Pratt, P. L. (1964). Stress relaxation and the plastic deformation of solids. *Phys. Stat. Sol.*, **6**, 111–20.

Guyot, P. (1968). On the mechanisms of deformation in dispersion hardened alloys. In *Oxide dispersion strengthening*, G. S. Ansell, T. D. Cooper & F. V. Lenel Eds. Gordon & Breach, New York, 405–27.

Guyot, P., & Dorn, J. E. (1967). A critical review of the Peierls mechanism. *Can. J. Phys.*, **45**, 983–1016.

Haasen, P., Davis, L. A., Aladag, E., & Gordon, R. B. (1970). On the mechanism of stage III deformation in NaCl single crystals. *Scripta Metall.*, **4**, 55–6.

Haasen, P., & Lawson, A. W. (1958). Der Einfluß hydrostatischen Druckes auf die Zugverformung von Einkristallen. *Zeit. für Metallkunde*, **49**, 280–91.

Hall, E. L., Walter, J. L., & Briant, C. L. (1982). The structure of [110] tilt boundaries in polycrystalline Fe 3% Si. *Philos. Mag.*, A.**45**, 753–70.

Hardwick, D., Sellars, C. M., & Tegart, W. J. McG. (1961). The occurrence of recrystallization during high temperature creep. *J. Inst. Met.*, **90**, 21–2.

Harper, J., & Dorn, J. E. (1957). Viscous creep of aluminium near its melting temperature. *Acta Metall.*, **5**, 654–65.

Harris, J. E., & Jones, R. B. (1963). Directional diffusion in magnesium alloys. *J. Nuclear Mater.*, **10**, 360–2.

Hart, E. W. (1967). Theory of the tensile test. *Acta Metall.*, **15**, 351–5.

Hart, E. W. (1970). A phenomenological theory for plastic deformation of polycrystalline metals. *Acta Metall.*, **18**, 599–610.

Hart, E. W. (1976). Constitutive relations for the nonelastic deformation of metals. *J. Eng. Materials & Technology*, **98**, 193–202.

Hart, E. W. (1981). A critical examination of steady-state creep. In *Creep in structures*, A. R. S. Ponter & D. R. Hayhurst Eds. Springer-Verlag, Berlin.

Hart, E. W., & Solomon, H. D. (1973). Load relaxation studies of polycrystalline high purity aluminium. *Acta Metall.*, **21**, 295–307.

Hasegawa, T., Karashima, S., & Hasegawa, R. (1971). Substructure formation and non-uniformity in strain during high temperature creep of Copper single crystals. *Metall. Trans.*, **2**, 1449–55.

Haworth, W. L., & Gordon, R. B. (1970). Pressure dependence of dislocation mobility in ionic crystals. *Phys. Stat. Sol. (a)*, **3**, 503–10.

Hayden, H. W., Gibson, R. C., Merrick, H. F., & Brophy, J. F. (1967). Superplasticity in the Ni–Fe–Cr system. *Trans. ASM*, **60**, 3–14.

Heard, H. C., & Kirby, S. H. (1981). Activation volume for steady-state creep in polycrystalline CsCl: cesium chloride structure. In *Mechanical behavior of crustal rocks*, Geophysical Monograph 24, American Geophysical Union, 83—91.

Herring, C. (1950). Diffusional viscosity of a polycrystalline solid. *J. Appl. Phys.*, **21**, 437–45.

Herring, C. (1951). Surface tension as a motivation for sintering. In *The physics of powder metallurgy*, W. E. Kingston Ed. McGraw-Hill, New York, ch. 8.

Hillert, M., & Purdy, G. R. (1978). Chemically induced grain boundary migration. *Acta Metall.*, **26**, 333–40.

Hirth, J. P. (1970). Thermodynamics of stacking-faults. *Met. Trans.*, **1**, 2367–74.

Hirth, J. P. (1972). The influence of grain-boundaries on mechanical properties. *Metall. Trans.*, **3**, 3047–67.

Hirth, J. P., & Balluffi, R. W. (1973). On grain-boundary dislocation and ledges. *Acta Metall.*, **21**, 929–42.

Hirth, J. P., & Nix, W. D. (1969). An analysis of the thermodynamics of dislocation glide. *Phys. Stat. Sol.*, **35**, 177–88.

Hirthe, W. M., & Brittain, J. O. (1963). High-temperature steady-state creep of rutile. *J. Amer. Ceram. Soc.*, **46**, 411–17.

Hobbs, B. E. (1968). Recrystallization of single crystals of quartz. *Tectonophysics*, **6**, 353–401.

Hobbs, B. E., McLaren, A. C., & Paterson, M. S. (1972). Plasticity of single crystals of synthetic quartz. In *Flow and fracture of rocks*, H. C. Heard, I. Y. Borg, N. C. Carter & C. B. Raleigh Eds. AGU, Washington, D.C., 29–53.

Hornstra, J. (1960). Dislocation, stacking faults and twins in the spinel structure. *J. Phys. Chem. Solids*, **15**, 311–23.

Hüther, W., & Reppich, B. (1973). Dislocation structure during creep of MgO single crystals. *Philos. Mag.*, **28**, 363–71.

Immarigeon, J.-P., & Jonas, J. J. (1971). Flow stress and substructural change during the transient deformation of Armco iron and silicon steel. *Acta Metall.*, **19**, 1053–61.

Immarigeon, J.-P., & Jonas, J. J. (1974). The deformation of Armco iron and Silicon steel in the vicinity of the Curie temperature. *Acta Metall.*, **22**, 1235–47.

Inokuti, Y., & Doherty, R. D. (1978). Transmission Kossel study of the structure of compressed iron and its recrystallization behaviour. *Acta Metall.*, **26**, 61–80.

Ion, S. E., Humphreys, F. J., & White, S. H. (1982). Dynamic recrystallization and the development of microstructure during the high-temperature deformation of magnesium. *Acta Metall.*, **30**, 1909–19.

Jackson, J. W., & Waxman, M. (1963). An analysis of pressure and stress distribution under rigid Bridgman-type anvils. In *High-pressure measurement*, A. H. Giardini & E. C. Lloyd Eds. Butterworths, London, 39–48.

Jaoul, O., & Houlier, B. (1983). Study of ^{18}O diffusion in magnesium orthosilicate by nuclear micro analysis. *J. Geophys. Res.*, **88**, 613–24.

Jaoul, O., Poumellec, M., Froidevaux, C., & Havette, A. (1981). Silicon diffusion in forsterite: A new constraint for understanding mantle deformation. In *Anelasticity in the Earth*. Geodynamics series, 4, AGU.

Jeanloz, R., & Richter, F. M. (1979). Convection, composition and the thermal state of the lower mantle. *J. Geophys. Res.*, **84**, 5497–503.

Jenkins, C. H. M., & Mellor, G. A. (1935). Investigation of the behaviours of metals under deformation at high temperature. I – Structural changes in mild steel and commercial iron during creep. *J. Iron Steel Inst.*, **132**, 179–227.

Jesser, W. A., & Kuhlmann-Wilsdorf, D. (1972). The flow stress and dislocation structure of nickel deformed at very high pressure. *Mat. Sci. & Energ.*, **9**, 111–17.

Johnson, C. A., Bradt, R. C., & Hoke, J. H. (1975). Transformational plasticity in Bi_2O_3. *J. Amer. Ceram. Soc.*, **58**, 37–40.

Jonas, J. J. (1969). A comparison of creep and hot-working strain-rate relationships. *Trans. A.S.M.*, **62**, 300–3.

Jonas, J. J., Holt, R. A., & Coleman, C. E. (1976). Plastic stability in tension and compression. *Acta Metall.*, **24**, 911–18.

Jonas, J. J., & Luton, M. J. (1978). Flow softening at elevated temperatures. In *Advances in deformation processing*, J. J. Burke & V. Weiss Eds. Plenum, New York, 596 pp.

Jonas, J. J., Sellars, C. M., & Tegart, W. J. McG. (1969). Strength and structure under hot working conditions. *Metall. Rev.*, **130**, 1–24.

Jones, H. (1969). A comparison of theory with experiment for diffusion creep in metal foils and wires. *Mater. Sci. & Eng.*, **4**, 106–14.

Jones, S. J., & Glen, J. W. (1969). The effect of dissolved impurities on the mechanical properties of ice crystals. *Philos. Mag.*, **19**, 13–24.

Kamb, W. B. (1961). The thermodynamic theory of non-hydrostatically stressed solids. *J. Geophys. Res.*, **66**, 259–71.

Kamb, W. B. (1973). Theory of preferred orientation developed by crystallization under stress. *J. Geophys.*, **67**, 153–70.

Karato, S. (1981a). Rheology of the lower mantle. *Phys. Earth Planet. Int.*, **24**, 1–14.

Karato, S. (1981b). Pressure dependence of diffusion in ionic solids. *Phys. Earth Planet. Int.*, **25**, 38–51.

Karato, S., Toriumi, M., & Fujii, T. (1980). Dynamic recrystallization of olivine single crystals during high temperature creep. *Geophys. Res. Lett.*, **7**, 649–52.

Karato, S., Toriumi, M., & Fujii, T. (1982). Dynamic recrystallization and high-temperature rheology of olivine. In *High pressure research in geophysics*, S. Akimoto & M. Manghnani Eds. Center for Academic Publications, Tokyo, 171–89.

Kekulawala, K. R. S. S., Paterson, M. S., & Boland, J. N. (1978). Hydrolytic weakening in quarts. *Tectonophys.*, **46**, T1–6.

Kekulawala, K. R. S. S., Paterson, M. S., & Boland, J. N. (1981). An experimental study of the role of water in quartz deformation. In *Mechanical behaviour of crystal rocks*. Geophysical Monograph 24, AGU.

Kennedy, A. J. (1962). *Process of creep and fatigue in metals*. Oliver & Boyd, London, 480 pp.

Keyes, R. W. (1958). Volumes of activation for diffusion in solids. *J. Chem. Phys.*, **29**, 467–75.

Keyes, R. W. (1963). Continuum models of the effect of pressure on activated processes. In *Solids under pressure*, W. Paul & D. M. Warschauer Eds. McGraw-Hill, New York.

Kohlstedt, D. L., & Goetze, C. (1974). Low-stress, high-temperature creep in olivine single crystals. *J. Geophys. Res.*, **79**, 2045–51.

Kohlstedt, D. L., Goetze, C., Durham, W. B., & van der Sande, J. B. (1976). A new technique for decorating dislocations in olivine. *Science*, **191**, 1045–6.

Kohlstedt, D. L., & Weathers, M. S. (1980). Deformation-induced microstructures, paleo-piezometers, and differential stresses in deeply eroded fault zones. *J. Geophys. Res.*, **85**, 6269–85.

Kot, R. A., & Weiss, V. (1970). Transformation plasticity in iron–nickel alloys. *Metall. Trans.*, **1**, 2685–93.

Krivanek, O. L., Isoda, S., & Kobayashi, K. (1977). Lattice imaging of a grain-boundary in crystalline germanium. *Philos. Mag.*, **36**, 931–40.

Kronberg, M. L. (1957). Plastic deformation of single crystals of sapphire: Basal slip and twinning. *Acta Metall.*, **5**, 507–23.

Kroupa, F., & Vitek, V. (1964). Splitting of dislocations in bcc metals on {100} planes. *Czech. J. Phys.*, B.**14**, 337–46.

Lacroix, A. (1903). Observations sur la cristallisation du zinc par recuit, faites dans les ruines incendiées de Saint-Pierre (Martinique). *Bull. Soc. Fr. Minéral.*, **26**, 184–8.

Langdon, T. G., & Mohamed, F. A. (1977). The characteristics of independent and sequential creep processes. *J. Austr. Inst. Met.*, **22**, 189–99.

Langdon, T. G., & Yavari, P. (1982). An investigation of Harper–Dorn creep. II – The flow process. *Acta Metall.*, **30**, 881–7.

Lasaga, A. C. (1979). Multicomponent exchange and diffusion in silicates. *Geochim. & Cosmochim. Acta*, **43**, 455–9.

le Hazif, R., & Poirier, J. P. (1975). Cross-slip on {110} planes in aluminium single crystals compressed along ⟨100⟩ axis. *Acta Metall.*, **23**, 865–71.

Levade, C., Couderc, J. J., Bras, J., & Fagot, M. (1982). Transmission electron microscopy study of experimentally deformed pyrite. *Philos. Mag.*, A.**46**, 307–25.

Lifshitz, I. M. (1963). On the theory of diffusion-viscous flow of polycrystalline bodies. *Soviet Physics JETP*, **17**, 909–20.

Liu, L. G. (1979). Phase transformations and the constitution of the deep mantle. In *The Earth: its origin, structure and evolution*, M. W. McElhinny Ed. Academic Press, London, 177–202.

Lücke, K., & Stüwe, N. P. (1971). On the theory of impurity controlled grain boundary motion. *Acta Metall.*, **19**, 1087–99.

Luton, M. J., & Sellars, C. M. (1969). Dynamic recrystallization in nickel and nickel iron alloys during high temperature deformation. *Acta Metall.*, **17**, 1033–43.

Mackwell, S. J., & Paterson, M. S. (1983). Water-related diffusion and deformation effects in quartz at pressures of 1500 and 300 MPa. *Fallen Leaf Chapman Conference on point defects in minerals.*

Madon, M., & Poirier, J. P. (1980). Dislocation in spinel and garnet high-pressure polymorphs of olivine and pyroxene: Implications for mantle rheology. *Science*, **207**, 66–8.

Maï, C., Perez, J., Rivière, R., Tatibouet, J., & Vassoille, R. (1977). Dislocations et propriété physico-mécaniques de la glace I_h. *Annales de Physique*, **2**, 91–118.

Martin, G., & Perraillon, B. (1980). Measurements on grain-boundary diffusion. In *Grain-boundary structure and kinetics*. American Society for Metals, Metals Park, Ohio, 470 pp.

Marya, S. K., & Wyon, G. (1975). Superplasticité à l'ambiante de l'aluminium à grain fin, en liaison avec l'existence d'un film intergranulaire de solution solide riche en gallium. *J. Physique*, **36**, C. 4, 309–13.

Maurice, F., Zemskoff, A., Dorizzi, P., & Poirier, J. P. (1973). Etude de la polygonisation de fluage au moyen des images électroniques au microanalyseur à sonde électronique. *Micron*, **4**, 184–92.

McCartney, L. N. (1976). No time – gentlemen please! *Philos. Mag.*, **33**, 689–95.

McClay, K. R. (1977). Pressure solution and Coble creep in rocks and minerals: a review. *J. Geol. Soc. Lond.*, **134**, 57–70.

McClay, K. R., & Atkinson, B. K. (1977). Experimentally induced kinking and annealing of single crystals of galena. *Tectonophysics*, **39**, 175–89.

McCormick, P. G., & Ruoff, A. L. (1969). Hydrostatic pressure and the mechanism of creep in aluminium. *J. Appl. Phys.*, **40**, 4812–18.

McKenzie, D. P. (1967). The viscosity of the mantle. *Geophys. J. R. Astr. Soc.*, **14**, 297–305.

McLaren, A. C., Cook, R. F., Hyde, S. T., & Tobin, R. C. (1983). The mechanisms of the formation and growth of water bubbles and associated dislocation loops in synthetic quartz. *Phys. Chem. Minerals*, **9**, 79–94.

McLaren, A. C., & Retchford, J. A. (1969). Transmission electron microscope study of the dislocations in plastically deformed quartz. *Phys. Stat. Sol.*, **33**, 657–68.

McLean, D. (1952). Crystal fragmentation in aluminium during creep. *J. Inst. Metals*, **81**, 287–92.

McLean, D. (1966). The physics of high-temperature creep in metals. *Rep. on Progress in Physics*, **29**, 1–33.

Means, W. D. (1983). Microstructure and micromotion in recrystallization flow of octachloropropane: a first look. In *Sander Volume, Geol. Rundschau* (in press).

Meisel, L. V., & Cote, P. J. (1983). Non-isothermal transformation kinetics: application to metastable phases. *Acta Metall.*, **31**, 1053–9.

248   References

Mendelson, S. (1961). Dislocation etch pit formation in sodium chloride. *J. Appl. Phys.*, **32**, 1579–83.

Mercier, J. C. C. (1980). Magnitude of the continental lithospheric stresses inferred from rheomorphic petrology. *J. Geophys. Res.*, **85**, 6293–303.

Mercier, J. C. C., Anderson, D. A., & Carter, N. L. (1977). Stress in the lithosphere: inference from the steady state flow or rocks. *Pageoph.*, **115**, 199–226.

Miller, A. K., Robinson, S. L., & Sherby, O. D. (1977). On the importance of subgrain size to the creep rate. *Philos. Mag.*, **36**, 757–60.

Mitchell, T. E., Hobbs, L. W., Heuer, A. H., Castaing, J., Cadoz, J., & Philibert, J. (1979). Interaction between point defects and dislocations in oxides. *Acta Metall.*, **27**, 1677–91.

Mitchell, T. E., Pletka, B. J., Philips, D. S., & Heuer, A. H. (1976). Climb dissociation of dislocations in sapphire (α-Al_2O_3). *Philos. Mag.*, **34**, 441–51.

Mitra, S. K., & McLean (1966). Work-hardening and recovery in creep. *Proc. Roy. Soc.*, A.**295**, 288–99.

Mitra, S. K., & McLean (1967). Cold work and recovery in creep at ostensibly constant structure. *Metal Sci. J.*, **1**, 192–8.

Mohamed, F. A., & Ginter, T. J. (1982). On the nature and origin of Harper–Dorn creep. *Acta Metall.*, **30**, 1869–81.

Mohamed, F. A., & Langdon, T. G. (1974). The transition from dislocation climb to viscous glide in creep of solid solution alloys. *Acta Metall.*, **22**, 779–88.

Mohamed, F. A,, & Langdon, T. G. (1976). The determination of activation energy for superplastic flow. *Phys. Stat. Sol. (a)*, **33**, 375–81.

Mukherjee, A. K., Bird, J. E., & Dorn, J. E. (1969). Experimental correlations for high-temperature creep. *Trans. A.S.M.*, **62**, 155–79.

Myshlyaev, M. M. (1976). Creep and dislocation structure of crystals at moderate temperatures. *Proceedings 4th ICSMA, ENSMIM, Nancy*, 1037–85.

Myshlyaev, M. M., Nikitenko, V. I., & Nesterenko, V. I. (1969). Dislocation structure and macroscopic characteristics of plastic deformation at creep of silicon crystals. *Phys. Stat. Sol.*, **36**, 89–96.

Nabarro, F. R. N. (1948). Deformation of crystals by the motion of single ions. *Report of a Conference on Strength of Solids (Bristol)*. The Physical Soc., 75–90.

Nabarro, F. R. N. (1967). Steady state diffusional creep. *Philos. Mag.*, **16**, 231–7.

Naziri, H., Pearce, R., Henderson-Brown, M., & Hale, K. F. (1973). In situ superplasticity experiments in the 1 Million Volt electron microscope. *J. of Microscopy*, **97**, 229–38.

Naziri, H., Pearce, R., Henderson-Brown, M., & Hale, K. F. (1975). Microstructural-mechanism relationship in the zinc–aluminium entectoid superplastic alloy. *Acta Metall.*, **23**, 489–96.

Nicolas, A., & Poirier, J. P. (1976). *Crystalline plasticity and solid state flow in metamorphic rocks*. Wiley Interscience, London, 444 pp.

Nuttall, K. (1971). Strain-rate sensitivity in superplastic metals. *Int. J. Mech. Sci.*, **13**, 83.

Nye, J. F. (1953). Some geometrical relations in dislocated crystals. *Acta Metall.*, **1**, 153–62.

Nye, J. F. (1957). *Physical properties of crystals.* Clarendon Press, Oxford, 322 pp.

O'Connell, R. J. (1977). On the scale of mantle convection. *Tectonophysics*, **38**, 119–36.

Oelschlägel, D., & Weiss, V. (1966). Superplasticity of steels during the ferrite–austenite transformation. *Trans. A.S.M.*, **59**, 143–54.

Oguchi, A., Yoshida, S., & Nobuki, M. (1972*a*). Influence of high hydrostatic pressure on the flow stress of zinc and zirconium polycrystals. *Trans. Japan Inst. Metals*, **13**, 69–75.

Oguchi, A., Yoshida, S., & Nobuki, M. (1972*b*). Influence of high hydrostatic pressure on the flow stress of pure iron and molybdenum polycrystals. *Trans. Japan Inst. Metals.*, **13**, 63–8.

Orlova, A., Pahutova, M., & Cadek, J. (1971). Dislocation structure and the effect of stacking-fault energy in high-temperature creep of Cu–Al solid solutions. *Philos. Mag.*, **23**, 303–12.

Orowan, E. (1940). Problems of plastic gliding. *Proc. Phys. Soc.*, **52**, 8–22.

Padmanabhan, K. A., & Davies, J. G. (1980). *Superplasticity.* Springer-Verlag, Berlin, 312 pp.

Parker, J. D., & Wilshire, B. (1976). On the subgrain size dependence of creep. *Philos. Mag.*, **34**, 485–9.

Parmentier, E. M. (1981). A possible mantle instability due to superplastic deformation associated with phase transitions. *Geophys. Res. Lett.*, **8**, 143–6.

Parsonage, N. G., & Staveley, L. A. K. (1978). *Disorder in crystals.* Clarendon Press, Oxford, ch. 8.

Paterson, M. S. (1963). Secondary changes of length with pressure in experimentally deformed rocks. *Proc. Roy. Soc. London*, A.**271**, 57–87.

Paterson, M. S. (1973). Non-hydrostatic thermodynamics and its geologic applications. *Rev. Geophys. Space Phys.*, **11**, 355–89.

Paterson, M. S. (1978). *Experimental rock deformation. The brittle field.* Springer-Verlag, Berlin, 254 pp.

Paterson, M. S. (1983). Creep in transforming polycrystalline materials. *Mechanics of Materials*, **2**, 103–9.

Paterson, M. S. (19xx). *Experimental rock deformation. The ductile field* (in preparation).

Paterson, M. S., & Kekulawala, K. R. S. S. (1979). The rôle of water in quartz deformation. *Bull. Minéral.*, **102**, 92–8.

Pearson, C. E. (1934). The viscous properties of extruded eutectic alloys of lead–tin and bismuth–tin. *J. Inst. Met.*, **54**, 111–24.

Peltier, W. R., & Jarvis, G. T. (1982). Whole mantle convection and the thermal evolution of the Earth. *Phys. Earth Planet. Int.*, **29**, 281–304.

Perdrix, C., Perrin, M. Y., & Montheillet, F. (1981). Comportement mécanique et évolution structurale de l'aluminium au cours d'une déformation à chaud de grande amplitude. *Mem. Sci. Rev. Metallurgie*, **78**, 309–20.

Phakey, P., Dollinger, G., & Christie, J. (1972). Transmission electron microscopy of experimentally deformed olivine crystals. In *Flow and fracture of rocks* (Griggs volume), H. C. Heard, I. Y. Borg, N. L. Carter & C. B. Raleigh Eds. AGU, Washington, 1972, 112–38.

Phillips, D. S., & Cadoz, J. L. (1982). Climb dissociation of ⟨1010⟩ dislocations in sapphire (α-Al_2O_3). *Philos. Mag.*, A.**46**, 583–95.

Poirier, J. P. (1972). High temperature creep of single crystalline sodium chloride. *Philos. Mag.*, **26**, 701–12, 713–25.

Poirier, J. P. (1975). On the slip systems of olivine. *J. Geophys. Res.*, **80**, 4059–61.

Poirier, J. P. (1976). On the symmetrical role of cross-slip of screw dislocations and climb of edge dislocations as recovery processes controlling high temperature creep. *Rev. Phys. Appl.*, **11**, 731–8.

Poirier, J. P. (1977). Microscopic creep models and the interpretation of stress-drop tests during creep. *Acta Metall.*, **25**, 913–17.

Poirier, J. P. (1978). Is power-law creep diffusion controlled? *Acta Metall.*, **26**, 629–37.

Poirier, J. P. (1979). Reply to 'Diffusion-controlled dislocation creep, a defense'. *Acta Metall.*, **27**, 401–3.

Poirier, J. P. (1980). Shear localization and shear instability in the ductile field. *J. Struct. Geol.*, **2**, 135–42.

Poirier, J. P. (1981). On the kinetics of olivine spinel transition. *Phys. Earth Planet. Int.*, **26**, 179–87.

Poirier, J. P. (1982). On transformation plasticity. *J. Geophys. Res.*, **87**, 6791–7.

Poirier, J. P., & Guillopé, M. (1979). Deformation-induced recrystallization of minerals. *Bull. Minéral.*, **102**, 67–74.

Poirier, J. P., & Nicolas, A. (1975). Deformation-induced recrystallization by progressive misorientation of subgrain-boundaries, with special reference to mantle peridotites. *J. Geology*, **83**, 707–20.

Poirier, J. P., Peyronneau, J., Gesland, J. Y., & Brebec, G. (1983). Viscosity and conductivity of the lower mantle; an experimental study on a $MgSiO_3$ perovskite analogue: $KZnF_3$. *Phys. Earth Planet. Int.*, **32**, 273–87.

Poirier, J. P., Sotin, C., & Peyronneau, J. (1981). Viscosity of high-pressure ice VI and evolution and dynamics of Ganymede. *Nature*, **292**, 225–7.

Poirier, J. P., & Vergobbi, B. (1978). Splitting of dislocation in olivine, cross-slip controlled creep and mantle rheology. *Phys. Earth Planet. Int.*, **16**, 370–8.

Pond, R. C., & Smith, D. A. (1977). On the absorption of dislocation by grain boundaries. *Philos. Mag.*, **36**, 353–66.

Pond, R. C., Smith, D. A., & Southerden, P. W. J. (1978). On the role of grain boundary dislocations in high-temperature creep. *Philos. Mag.*, A.**37**, 27–40.

Pontikis, V., & Poirier, J. P. (1975). Phenomenological and structural analysis of recovery-controlled creep, with special reference to the creep of single-crystal silver chloride. *Philos. Mag.*, **32**, 577–92.

Post, R. L. (1977). High temperature creep of Mt Burnett dunite. *Tectonophysics*, **42**, 75–110.

Presnyakov, A. A., & Chervyakova, V. V. (1959). K voprosu o 'Sverkhplastichnosti' splavov. *Fizika Metall. i Metalloved.*, **8**, 114–21.

Priester, L. (1980). Approche géométrique des joints de grains. Intérêt et limite. *Revue Phys. Appl.*, **15**, 789–830.

Putnis, A., & McConnell, J. D. C. (1980). *Principles of mineral behaviour.* Elsevier, New York, 257 pp.

Rae, C. M. F., & Smith, D. A. (1980). On the mechanisms of grain boundary migration. *Philos. Mag.*, A.**41**, 477–92.

Raj, R., & Ashby, M. F. (1971). On grain boundary sliding and diffusional creep. *Metall. Trans.*, **2**, 1113–27.

Raj, R., & Ghosh, A. K. (1981). Micromechanical modelling of creep using distributed parameters. *Acta Metall.*, **29**, 283–92.

Rao, C. N. R., & Rao, K. J. (1978). *Phase transitions in solids.* McGraw-Hill, New York, 330 pp.

Reiner, M. (1969). *Deformation, strain and flow.* H. K. Lewis, London, 347 pp.

Richardson, G. J., Sellars, C. M., & Tegart, W. J. McG. (1966). Recrystallization during creep of nickel. *Acta Metall.*, **14**, 1225–36.

Richter, F. M., & McKenzie, D. P. (1981). On some consequences and possible causes of layered mantle convection. *J. Geophys. Res.*, **86**, 6133–42.

Roberts, A. C., & Cottrell, A. H. (1956). Creep of alpha uranium during irradiation with neutrons. *Philos. Mag.*, **1**, 711–17.

Robinson, W. H. (1968). Dislocation etch pit techniques. In *Techniques of metals research*, R. F. Bunshah Ed. Interscience, New York, **2**, 291–340.

Rosenhain, W., Haughton, J. L., & Bingham, K. E. (1920). Zinc alloys with aluminium and copper. *J. Inst. Met.*, **23**, 261–323.

Ross, J. V., Ave Lallemant, H. G., & Carter, N. L. (1979). Activation volume for creep in the upper mantle. *Science*, **203**, 261–3.

Ross, J. V., Ave Lallemant, H. G., & Carter, N. L. (1980). Stress dependence of recrystallized grain and subgrain size in olivine. *Tectonophysics*, **70**, 39–61.

Rossard, C., & Blain, P. (1958). Premiers résultats de recherches sur la déformation des aciers à chaud. Mise au point d'un appareillage spécialement étudié. *Rev. Métallurgie*, **55**, 573–94.

Roy, R. (1973). A syncretist classification of phase transitions. In *Phase transitions 1973*, L. E. Cross Ed. Pergamon Press, Oxford.

Ruoff, A. L. (1965). Mass transfer problems in ionic crystals with charge neutrality. *J. Appl. Phys.*, **36**, 2903–7.

Rutter, E. H., Atkinson, B. K., & Mainprice, D. H. (1978). On the use of the stress relaxation testing method in studies of the mechanical behaviour of geological materials. *Geophys. J. R. Astr. Soc.*, **55**, 155–70.

Sammis, C. G., & Dein, J. L. (1974). On the possibility of transformational superplasticity in the Earth's mantle. *J. Geophys. Res.*, **79**, 2961–5.

Sammis, C. G., Smith, J. C., & Schubert, G. (1981). A critical assessment of estimation methods for activation volume. *J. Geophys. Res.*, **86**, 10707–18.

Sammis, C. G., Smith, J. C., Schubert, G., & Yuen, D. A. (1977). Viscosity depth profile of the Earth's mantle. Effect of polymorphic phase transitions. *J. Geophys. Res.*, **82**, 3747–61.

Sauveur, A. (1924). What is Steel? Another answer. *The Iron Age*, **113**, 581–3.

Schmid, E., & Boas, W. (1950). *Plasticity of crystals.* Hughes, London.

Schmid, S., Boland, J. N., & Paterson, M. S. (1977). Superplastic flow in fine grained limestone. *Tectonophysics*, **43**, 257–91.

Schmid, S. M., Paterson, M. S., & Boland, J. N. (1980). High temperature flow and dynamic recrystallization in Carrara marble. *Tectonophysics*, **65**, 245–80.

Seeger, A., & Haasen, P. (1958). Density changes in crystals containing dislocations. *Philos. Mag.*, **3**, 470–5.

Sellars, C. M. (1978). Recrystallization of metals during hot deformation. *Phil. Trans. R. Soc. London*, A.**288**, 147–58.

Sellars, C. M., & Quarrell, A. G. (1961). The high-temperature creep of gold-nickel alloys. *J. Inst. Metals*, **90**, 329–36.

Sherby, O. D., & Burke, P. M. (1967). Mechanical behaviour of crystalline solids at elevated temperature. *Progr. Mat. Sci.*, **13**, 325–90.

Sherby, O. D., Klundt, R. H., & Miller, A. L. (1977). Flow-stress, subgrain size and subgrain stability at elevated temperatures. *Metall. Trans.*, **8A**, 843–50.

Sherby, O. D., Robbins, J. L., & Goldberg, A. (1970). Calculation of activation volumes for self-diffusion and creep at high temperatures. *J. Appl. Phys.*, **41**, 3961–8.

Sherby, O. D., & Weertman, J. (1979). Diffusion-controlled dislocation creep, a defense. *Acta Metall.*, **27**, 387–400.

Shewmon, P. G. (1981). Diffusion driven grain boundary migration. *Acta Metall.*, **29**, 1567–72.

Siethoff, H. (1983). Cross-slip in the high temperature deformation of germanium, silicon and indium antimonide. *Philos. Mag.*, A.**47**, 657–69.

Smidoda, K., Gottschalk, W., & Gleiter, H. (1978). Diffusion in migrating interfaces. *Acta Metall.*, **26**, 1833–6.

Smith, J. V. (1982). *Geometrical and structural crystallography*. Wiley, New York, 450 pp.

Smith, D. A., Rae, C. M. F., & Grovenor, C. R. M. (1979). Grain-boundary migration. In *Grain boundary structure and kinetics*. ASM, Metals Park, Ohio, 470 pp.

Smyth, D. M., & Stocker, R. L. (1975). Point defects and non-stoichiometry in forsterite. *Phys. Earth Planet. Inter.*, **10**, 183–92.

Smyth, J. R., Bradt, R. C., & Hoke, J. H. (1975). Transformational superplasticity in the Bi_2O_3–Sm_2O_3 entectoid system. *J. Amer. Ceram. Soc.*, **58**, 381–4.

Solomon, A. A., & Nix, W. D. (1970). Interpretation of high temperature plastic deformation in terms of measured effective stress. *Acta Metall.*, **18**, 863–76.

Spencer, G. C. (1968). *Introduction to plasticity*. Chapman & Hall, London, 118 pp.

Spingarn, J. R., Barnett, D. M., & Nix, W. D. (1979). Theoretical descriptions of climb controlled steady state creep at high and intermediate temperatures. *Acta Metall.*, **27**, 1549–61.

Spohn, T., & Schubert, G. (1982). Models of mantle convection and the removal of heat from the Earth's interior. *J. Geophys. Res.*, **87**, 4682–96.

Squires, R. L., Weiner, R. T., & Philips, M. (1963). Grain boundary denuded zones in a magnesium 1/2 wt % zirconium alloy. *J. Nuclear Mater.*, **8**, 77–80.

Sritharan, T., & Jones, H. (1979). The creep of beta-cobalt at low stresses. *Acta Metal.*, **27**, 1293–1300.

Stevens, R. N. (1972). Grain boundary sliding and diffusion creep. *Surface Science*, **31**, 543–65.

Stocker, R. L., & Ashby, M. F. (1973*a*). On the rheology of the upper mantle. *Rev. Geophys. Space Phys.*, **11**, 391–426.

Stocker, R. L., & Ashby, M. F. (1973*b*). On the empirical constants in the Dorn equation. *Scripta Metall.*, **7**, 115–20.

Stohr, J. F., & Poirier, J. P. (1972). Etude en microscopie électronique du glissement pyramidal {1122} ⟨1123⟩ dans le magnésium. *Philos. Mag.*, **25**, 1313–29.

Strunk, H. (1975). Investigation of cross-slip events in NaCl crystals by transmission electron microscopy. *Phys. Stat. Sol. (a)*, **28**, 119–26.

Takeuchi, S., & Argon, A. S. (1976*a*). Steady-state creep of simple phase crystalline matter at high temperatures. *J. Mater. Sci.*, **11**, 1542–66.

Takeuchi, S., & Argon, A. S. (1976*b*). Steady-state creep of alloys due to viscous motion of dislocations. *Acta Metall.*, **24**, 883–9.

Takeuchi, S., & Argon, A. S. (1979). Glide and climb resistance to the motion of an edge dislocation due to dragging a Cottrell atmosphere. *Philos. Mag.*, A.**40**, 65–75.

Taplin, D. M. R., Dunlop, G. L., & Langdon, T. G. (1979). Flow and failure of superplastic materials. *Ann. Rev. Mater. Sci.*, **9**, 151–89.

Tonejc, A., & Poirier, J. P. (1977). High temperature creep of equiaxed Cd-26.5 at % Zn eutectic in the superplastic regime. *Proc. 4th Int. Conf. on the strength of metals and alloys, Nancy*, 451–5.

Trouton & Rankine, A. O. (1904). On the stretching and torsion of lead wire beyond the elastic limit. *Philos. Mag.*, **8**, 538–56.

Tullis, J., Shelton, G. L., & Yund, R. A. (1979). Pressure dependence of rock strength: implications for hydrolytic weakening. *Bull. Mineral.*, **102**, 110–14.

Tungatt, P. D., & Humphreys, F. J. (1981). An *in-situ* optical investigation of the deformation behaviour of sodium nitrate – an analogue for calcite. *Tectonophysics.*, **78**, 661–75.

Twiss, R. J. (1976). Structural superplastic creep and linear viscosity in the Earth's mantle. *Earth Planet. Sci. Lett.*, **33**, 86–100.

Twiss, R. J. (1977). Theory and applicability of a recrystallized grain size paleopiezometer. *Pageoph.*, **115**, 227–44.

Underwood, E. E. (1962). A review of superplasticity and related phenomena. *J. of Metals*, Dec., 914–19.

Urai, J. (1983). Water assisted dynamic recrystallization and weakening in polycrystalline bischofite. *Tectonophysics*, **96**, 125–57.

Urai, J. L., Humphreys, F. J., & Burrows, S. E. (1980). *In-situ* studies of the deformation and dynamic recrystallization of rhombohedral camphor. *J. Mater. Sci.*, **15**, 1231–40.

Uyeda, S. (1978). *The new view of the Earth.* W. H. Freeman, San Francisco, 217 pp.

Vagarali, S. S., & Langdon, T. G. (1981). Deformation mechanisms in hcp metals at elevated temperatures. I. Creep behavior of magnesium. *Acta Metall.*, **29**, 1969–82.

Verne, J. (1864). *Voyage au centre de la terre*, Hetzel, Paris.

Veyssière, P., Rabier, J., Garem, H., & Grilhe, J. (1978). Influence of temperature on dissociation of dislocations and plastic deformation in spinel oxides. *Philos. Mag.*, A.**38**, 61–79.

Veyssière, P., Rabier, J., & Grilhe, J. (1975). Stacking fault energy computations in oxides with normal and inverse spinel structures. *Phys. Stat. Sol.* (a), **31**, 605–14.

Vitek, V. (1974). Theory of the core structures of dislocations in body-centred-cubic metals. *Crystal Lattice Defects*, **5**, 1–34.

Vitek, V., Sutton, A. P., Smith, D. A., & Pond, R. C. (1979). Atomistic studies of grain boundaries and grain boundary dislocations. In *Grain boundary structure and kinetics*. ASM, Metals Park, Ohio, 470 pp.

Wassermann, G. (1937). Untersuchungen an einer Eisen–Nickel Legierung über die Verformbarkeit während der γ–α Unwandlung. *Archiv Für der Eisenhütt.*, **7**, 321–5.

Weertman, J. (1957). Steady-state creep of crystals. *J. Appl. Phys.*, **28**, 1185–9.

Weertman, J. (1960). Creep of indium, lead and some of their alloys with various metals. *Trans. AIME*, **218**, 207–18.

Weertman, J. (1968). Dislocation climb theory of steady-state creep. *Trans. A.S.M.*, **61**, 681–94.

Weertman, J. (1970). The creep strength of the Earth's mantle. *Rev. Geophys. Space Phys.*, **8**, 145–68.

Weertman, J. (1972). High temperature creep produced by dislocation motion. In *J. E. Dorn Memorial Symposium, Cleveland, Ohio, Oct. 17, 1972.*

Weertman, J. (1973). Creep of ice. In *Physics and chemistry of ice*, E. Whalley, S. J. Jones & L. W. Gold Eds. Roy. Soc. Canada, 320–37.

Weertman, J. (1978). Creep laws for the mantle of the Earth. *Phil. Trans. R. Soc. London*, A.**288**, 9–26.

Weertman, J. (1983). Creep deformation of ice. *Annual Review of Earth and Planetary Science*, **11**, 215–40.

Weertman, J., & Weertman, J. R. (1975). High temperature creep of rocks and mantle viscosity. *Ann. Rev. Earth Planet. Sci.*, **3**, 293–315.

Wegner, M., & Christie, J. M. (1974). Preferential chemical etching of terrestrial and lunar olivines. *Contr. Mineral. Petrol.*, **43**, 195–212.

Wegner, M., & Christie, J. M. (1983). Chemical etching of deformation substructures in Quartz. *Phys. Chem. Minerals*, **9**, 67–78.

Weissman, S., & Kalman, Z. H. (1969). X-Ray diffraction topographic methods. In *Techniques of metal research*, R. F. Bunshah Ed. Interscience, New York, **2**, 839–73.

White, S. (1973). Syntectonic recrystallization and texture development in quartz. *Nature*, **244**, 276–8.

White, S. H. (1976). The effects of strain and microstructure fabrics and deformation mechanisms in quartzite. *Philos. Trans. R. Soc. London*, A.**283**, 69–86.

White, S. H. (1979). Difficulties associated with paelostress estimates. *Bull. Minéral.*, **102**, 210–15.

Whitworth, R. W. (1978). The core structure and the mobility of dislocations in ice. *J. Glaciology*, **21**, 341–59.

Whitworth, R. W. (1980). The influence of the choice of glide plane on the theory of the velocity of dislocations in ice. *Philos. Mag.*, A.**41**, 521–8.

Whitworth, R. W. (1983). The velocity of dislocation in ice on {0001} and {1010} planes. *J. Phys. Chem.*, **87**, 000–00.

Winger, L. A., Bradt, R. C., & Hoke, J. H. (1980). Transformational superplasticity of Bi_2WO_6 and Bi_2MoO_6. *J. Amer. Ceram. Soc.*, **63**, 291–4.

Woodcock, A., & Davis, M. (1978). *Catastrophe theory*. E. P. Dutton, New York, 152 pp.

Wyon, G., & Crussard, C. (1951). Modifications de structure de l'aluminium au cours du fluage. *Rev. Metall.*, **48**, 121–30.

Yavari, P., Miller, D. A., & Langdon, T. G. (1982). An investigation of Harper–Dorn creep. I – Mechanical and microstructural characteristics. *Acta Metall.*, **30**, 871–9.

Yoshida, S., & Oguchi, A. (1970). Influence of high hydrostatic pressure on the flow stress of aluminium polycrystals. *Trans. Japan Inst. Metals*, **11**, 424–30.

Yu, E. C., & Li, J. C. M. (1977). Impression creep of LiF single crystals. *Philos. Mag.*, **36**, 811–25.

Yund, R. A., Smith, B. M., & Tullis, J. (1981). Dislocation-assisted diffusion in albite. *Phys. Chem. Minerals*, **7**, 185–9.

Zackay, V. F., Parker, E. R., Fahr, D., & Busch, R. (1967). The enhancement of ductility in high-strength steels. *Trans. A.S.M.*, **60**, 252–9.

Zamora, M., & Poirier, J. P. (1983). Experiments in anisothermal transformation plasticity: the case of cobalt. Geophysical implications. *Mechanics of Materials*, **2**, 193–202.

Zener, C., & Hollomon, J. H. (1944). Effect of strain rate upon plastic flow of steel. *J. Appl. Phys.*, **15**, 22–32.

Zeuch, D. H. (1983). On the inter-relationship between grain-size sensitive creep and dynamic recrystallization of olivine. *Tectonophysics*, **93**, 151–68.

Materials index

Subject index

Numbers in bold type refer to main entries within a group.